# Superbike 1990

## Gary Pinchin and Kel Edge

McKinnon Farmer Publishing Limited
33 Abbey Road, Grimsby, South Humberside, DN32 0HQ. Telephone 0472 361357. Fax 0472 241038

# Acknowledgements

The Editor would like to thank the following for their help in compiling Superbike 1990:

The Federation Internationale Motocycliste at their headquarters at Chambesy, Switzerland.
Their assistance with statistics and prompt details of date changes for 1990 was a great help.

In addition to his excellent photographic work, Kel Edge made himself responsible for collecting all race data throughout his travels in 1989. Without his help we could not have prepared such a detailed review of 1989.

Brian Catterson (Cycle News, USA) for his assistance in covering the Canadian and American rounds of the series.

While the publishers have taken all steps possible to make the statistics sections as comprehensive as they can, their ability to do so was limited to the amount of information contained on the results sheets provided by race organisers.

***PUBLISHER***

***McKinnon Farmer Publishing Limited***

*Published in March 1990 by McKinnon Farmer Publishing Limited,*
*33, Abbey Road, Grimsby, South Humberside, DN32 OHQ, England.*

***ISBN: 0 9514 711 3 9***

*Printed by Kim Hup Lee Printing Co Pte Limited, Singapore*

*Colour separations by Chroma Graphics(Overseas) Pte Limited, Singapore*

*Typesetting by Headlines (Grimsby) Limited, South Humberside, England.*

# Superbike 1990

# Contents

# The story so far...

For four years the loud Californian with pop star sunglasses lived out of a suitcase and travelled the world, hawking his product.

"This will save motorcycle racing!" he exclaimed as he seduced would-be financiers into backing his concept. The scepticism which surrounded many of his approaches has now subsided under an exciting wave of fact.

The fact is that Steve McLaughlin's sales pitch is fast becoming a proven fact...Superbike racing will doubtless emerge as motorcycle racing's saviour as grand prix competition nose-dives into an era of political unrest and technical evolution which is already eating its way into the once swollen 500cc grids.

There is no doubting the commercial success of grand prix racing, attendances have never been higher and multi-million dollar budgets for the factory teams have never been so abundant..it is truly an elitist sport where the rich will get richer and the poor not so rich.

What has happened in grand prix racing by and large is that the apprenticeship system which existed within the 500cc class has now disappeared.

With a few sensational exceptions, like John Kocinski, all future grand prix stars are likely to emerge from the ranks of four-stroke Superbike racing.

Superbike racing has preserved that vital element in road racing, every rider's right to prove himself within an arena where that is practically possible.

The rules governing Superbike racing are such that equal competition is maintained. OK, manufacturer superiority is beginning to creep in but as long as the sport's governing body, the Federation Internationale Motocycliste, keeps a tight rein on machine specifications the series will prosper.

It was perhaps understandable that the concept of a world championship for Superbikes should emerge from America. After all it was there that the premier national championship had been so successfully based on the Superbike class.

And it is worth remembering that names like Freddie Spencer, Eddie Lawson, Wayne Rainey and Kevin Schwantz all emerged from this domestic championship.

Early financiers of the World Superbike Championship concept were the France family of Daytona International Speedway fame. They helped McLaughlin make early ground in his campaign but then relinquished their interest.

But the backing which was to see the series come to fruition was to be supplied by the New Zealand-based Sports Marketing Company. Based in Auckland, the company was a subsidiary of Global Sports and Promotions which in turn was part of the publically quoted Madison Corporation.

McLaughlin got 35 per cent of the stock in The Sports Marketing Company for his idea and the partnership then embarked on a series of world tours to sign up circuit promoters and elusive sponsors.

The work in hand was to try and convince as many promoters as possible that getting involved in Superbike was a worthwhile proposition. Having already convinced the FIM that there was a need for another road racing world championship, McLaughlin's task was to present the sports governing body with a sufficient number of date applications for the 1988 season.

The Californian's job was made somewhat easier with the arrival in Europe of Keith Jones, Managing Director of both The Sports Marketing Company and its parent company.

A diminutive, hard-working character, Keith Jones perfectly complimented McLaughlin's powers of persuasion. He was a workaholic and perhaps most significantly obsessed by thorough administration.

So, an effective spearhead was formed to the Superbike movement which worked well during 1987 and presented a list of nine events to the FIM at its Paris Congress in October 1987.

The first major problem was therefore overcome. A world championship fixture list was in place for 1988 but what still evaded the project, and indeed was to prove its downfall some 18 months later, was a series sponsor to relieve The Sports Marketing Company Limited of the enormous operating costs which lay ahead.

The cost of keeping two executives in airline tickets and hotel rooms month after month presented Jones with an increasing

FM
YAMAHA
2
PLAYMOTO
FZR
TEAM
PIROVANO
YAMAHA
VALLI
POGGIPOLINI
YAMAHA
BYRD
Beta
Castrol
AFAM
MICHELIN

headache and this of course would be compounded by the bankroll required to run the races.

While the individual promoters were responsible for the prize schedule at their event, The Sports Marketing Company had made a big play on the fact that they would contribute £25,000 in travel assistance to riders at each event.

That meant that they would have to find a total of £225,000 in travelling expenses alone, before normal operating costs.

Despite the scary figures involved and the fact that the historic opening race at Donington Park arrived without any sign of a sponsor, the New Zealanders pressed on regardless, convinced that their product was good enough to attract big money without too much delay.

Right from the word go the series carried a hallmark of excitement. The crowds could see instantly that this was racing at its best with particularly satisfying crowds at some of the tracks.

Following the Donington Park round in Great Britain, the action moved to the impressive Hungaroring just outside Budapest and drew over 50,000 spectators.

The Hungarian Department of Tourism which had administered the event just could not believe their eyes. A similar attendance flocked to Sugo in Japan and perhaps most satisfying of all was the record-breaking attendance in Australia.

More than 20,000 people crammed themselves into Oran Park, near Sydney...far and away the biggest audience for a motorcycle event there.

Other events in the pioneer Superbike year, were staged in Austria, France, West Germany and New Zealand.

So, the enthusiasts had voted with their feet and given the World Superbike Championship a real vote of confidence as the opening season in this new category drew to a close.

Also satisfying was the level of manufacturer involvement in the series through importer teams. And although they didn't win the world title, the talking point of the championship had to be the performance of former 500cc World Champion, Marco Lucchinelli and the beautiful Ducati 851S.

Before the introduction of the World Superbike Championship Ducati and its fellow Italian producer, Bimota, had been languishing in the doldrums of an already receding motorcycle market.

But after just one season in competition their fortunes had considerably improved.

It seems that the correlation between riding motorcycles flat-out on the track and the High Street model had really captured the imagination of the big bike buyer and sales began to pick up significantly.

Not only was that true of the European manufacturer but also Japanese Superbike output was enhanced by this greater awareness of performance created by competition. Sales of models like the Honda RC30 became very bouyant.

The first year had to be termed an unqualified success. Fred Merkel took the very first World Superbike Championship through a consistent campaign against Italy's Davide Tardozzi on a Bimota.

But everything was not rosy behind the scenes. The major problem of operating without a sponsor still stared The Sports Marketing Company in the face. And with ever increasing debts, escalated by an incessant round of flights around the globe to try and find that vital backer, Keith Jones new that time was running out.

Through the winter of 1988 he would sometimes be on a different continent each day simply trying to make the hit which would bring him a richly deserved return for his confidence in the product.

Such was his determination that he kept appointments right until the week before the opening race in year two , again at Donington Park. Then he was called back to Auckland where a Madison Corporation board meeting decided to cut its losses and pull out of the

series.

Jones' relationship with McLaughlin had become more and more strained as the months had gone by and so several months earlier he had been ''bought out'' by The Sports Marketing Company.

Therefore it really was the end of an important chapter in the history of the World Superbike Championship. Love him or hate him, McLaughlin had been responsible for creating the new generation of motorcycle racing and Jones really had been the driving force behind the vital formative two years of the series(1987-1988).

There were few good words for the New Zealanders over Easter weekend, 1989. They had after all left the sport's governing body, the promoters and the riders high and dry with only a few days notice.

But if there is anything which can be said in defence of the ambitious Auckland-based company, it must be that without their time, effort and great expense we would not be enjoying the superb spectacle of World Superbike Championship racing today.

It cost The Sports Marketing Company Limited almost a million dollars for its flirtation with the project, a sum which they would surely have recouped had they stuck with it, but that was not to be.

In April 1989 the FIM had to quickly step in with financial assistance, in the form of the rider travel fund at Donington Park, to ensure continuity.

Almost immediately Japan showed an interest in the vacant World Superbike Championship. Using the UK offices of McKinnon Farmer, Tokyo-based Moto Co, specialists in motorcycle event promotions picked up the pieces of the crumbling series and using the British company for much of the early liaison between themselves and the FIM, started to put together a rescue plan.

Dentsu, the world's largest advertising agency, with interests, among a great many other things, in television production and distribution, began negotiations with the President of the FIM Road Race Commission, Joe Zegwaard, using Moto Co as their representatives.

It seemed certain that Dentsu would emerge as the sole stewards of the 1990 series but at the 11th hour European road race promoter, Maurizio Flammini stepped in with a rival bid which stole the contract from under the noses of the Japanese.

Subsequently a joint agreement was thrashed out between the Italian and Japanese companies with a new arrangement now in place for three years whereby the Flammini Group handles the day to day administration of the series and Dentsu the television production and distribution.

# Prospects

After stumbling its way through its first two seasons with little direction or security the World Superbike Championship has emerged with a seemingly exciting long term strategy.

Almost a year of protracted negotiations between Japan and Europe have thrashed out an intricate contract which awards ambitious promoter Maurizio Flammini the prime contract to run the series for the next three years and subsequently there has been a joint venture instigated between his Italian group and the Dentsu Corporation of Tokyo.

It was no easy ride and at times there were serious threats of rider boycotts which could have seriously jeopardised the future of the series. However the Superbike paddock is to be congratulated on its ultimate patience, which really formed the cornerstone of the rescue effort.

Without their continued support for the FIM, who found themselves in a very unenviable situation, there would have been no series about which to negotiate.

Those dismal days of uncertainty are now behind the riders and teams and they can look forward to a professionally promoted championship benefiting from global television. One of the major downfalls motorcycle racing has suffered in the past has been its failure to attract any organised television coverage. Grand prix racing has in recent years remedied that deficiency but air time for any other category has been negligible.

The arrival of Dentsu is indeed exciting. They are the world's best producers and distributors and their pedigree can be amply described by the fact that they have acquired the television rights for the 1992 Olympic

Games in Barcelona.

This is probably the major television event on earth and it is very encouraging that this same company will now be responsible for developing the image of Superbike racing throughout the world.

It will be Dentsu's responsibility to both produce footage of the races and distribute that film to as many broadcasters as possible. The sport must grow dramatically over the course of the next few years because of this fact.

The Japanese habitually take a long hard look at any project they have been invited to embark upon. The World Superbike Championship was one such challenge which they thought long and hard about and decided to accept. They are certain to succeed from now on.

The consequences of this major television development are very exciting to say the least. Vastly improved exposure will strengthen the financial base of the World Superbike Championship.

At present the series benefits from limited backing for major teams with similar commitment from manufacturers. The level of involvement in both these areas will grow dramatically as a result of Dentsu's pledge to put the series "on the air".

From the manufacturer's point of view the exploding television audience will represent a tremendous boost to their sales potential. In contrast to grand prix racing they are entering into competition machines, certainly identical in appearance, to those they are bolting together on their production lines.

This makes the promoting of those products far easier than with a two stroke grand prix machinery which only promotes brand awareness and not the specific qualities of that machine.

So the commercial undercurrent shows great potential and there is no doubt that this will be recognised by a growing number of industries.

The exciting potential of the series will also be recognised by major names in grand prix racing before too long. For 1990 Cabin Honda star Rob McElnea has joined the Loctite Yamaha World Superbike Championship team.

With a diminishing number of competitive 500cc grand prix machines the transfer market will hot up. Currently riders making the transition may feel that they are moving into the Second Division but when the effects of the new regime, and in particular Dentsu's power, begin to bite their feelings will alter.

The opportunities for earning a good living, comparable to those in the existing grand prix structure will multiply as the years go

***(Above)** Most striking newcomer of the World Superbike Championship scene was Giancarlo Falappa seen here in action on the Bimota. For 1990 he switches to Ducati power. **(Inset)** Terry Rymer carries Britain's hopes into the third year of Superbike.*

*Three of the biggest names in Superbike racing today.* ***(Right)*** *Belgian Stephane Mertens in pensive mood as he shelters from a startline shower* ***(far right, top)*** *Michael Dowson and minder* ***(below)*** *champion again, Fred Merkel.*

by.

Line-ups for 1990 already indicate that the upgrading of the series is beginning to take shape. Honda have always been the most strongly represented factory team in the series but the signs are that they will face tougher competition from Yamaha, Suzuki and Kawasaki this term.

Encouraged by their limited involvement last year Kawasaki will increase its presence through the Team Kawasaki Australia concern who are coming to Europe with Rob Phillis for an all-out attack on the 1990 title.

In only three 1989 outings Phillis amassed enough points to secure joint tenth position and the factory were obviously confident that they could bridge the performance gap during the course of the winter to have a meaningful crack at the title this year.

Partner Aaron Slight from New Zealand will also take part in selected rounds and who knows with former Yamaha star Michael Dowson now having joined the marque Kawasaki could really step up their effort later in the season if they felt the need!

Honda will once again rely on the skills of double World Superbike Champion Fred Merkel to try and make it a hat-trick of titles.The other mainstream Honda rider will be Belgian Stephane Mertens who ran so close last season.

One of the real threats again in 1990 will surely be tiny Italian Fabrizio Pirovano.For two seasons he has been one of the most consistent Superbike contenders and was robbed of any chance of success last time following an Oran Park crash which was not his fault.

However he is one of the most determined characters in world championship racing and will doubtless bounce back this year in search of Yamaha's first Superbike crown.

Sweden's Anders Andersson will also benefit from Yamaha factory involvement again. His fortunes were visibly enhanced by the switch to Yamaha from Suzuki for whom he had ridden all his career.

Japan will also be keeping a close eye on their British Yamaha entry, the Loctite Yamaha team.With the ever-improving Terry Rymer, who broke his duck in New Zealand last season, and former grand prix star Rob McElnea in their line-up, Yamaha must be excited by their prospects in the coming term.

Who would have thought,before the introduction of the World Superbike Championship in 1988,that we would have been tipping the Ducati factory to push for honours?But that will certainly be the case in 1990 with a new-look line-up which blends raw talent with experience.

Evergreen Raymond Roche will again lead the team but the sensational news is that Giancarlo Falappa, the spectacular former Bimota team member,has joined him in a partnership which will surely give the Japanese teams plenty to think about.

# Superbike 1990

**Round 1**
March 18
Jerez
Spain

**Round 2**
April 16
Donington Park
Great Britain

**Round 3**
April 29
Hungaroring
Hungary

**Round 4**
May 6
Hockenheim
West Germany

**Round 5**
May 20
Interlagos
Brazil

**Round 6**
June 3
Mosport Park
Canada

**Round 7**
June 10
Brainerd
USA

## Round 8

July 1

Osterreichring

Austria

## Round 9

August 26

Sugo

Japan

## Round 10

September 9

Le Mans

France

## Round 11

October 7

Monza

Italy

## Round 12

November 4

Shah Alam

Malaysia

## Round 13

November 11

Phillip Island

Australia

## Round 14

November 18

Manfeild

New Zealand

# Top Ten 1989

NOLAN
Dainese
UNDERWEAR
GIPSY
STO
ZUCCHINI

# 1
# Fred Merkel

In the two seasons of World Superbike, Fred Merkel has made the championship his own. Last year he was much more consistent than in 1988 thanks to better, HRC supplied equipment and a more settled team around him.

Backed by Italian businessman Oscar Rumi and Italian lingerie and tee shirt manufacturers, Gypsy, Merkel won three races last term and in the 22 races, only finished outside the top six on six occasions. He failed to score just once, at Sugo when he came home 16th in the first leg.

His choice of tyre company last term, Pirelli, was subject of many a derisive comment at the start of the year but Merkel got them working to his satisfaction and the tyres proved the match of any other manufacturer's products.

Only at the season opener did he strike problems when he was unable to use the compound he wanted, the new tyres were still under construction back in Italy. Even so he salvage fourth and sixth places and set himself up for the long season ahead.

Sugo was his low point of the year when he went 16-12 in the wet races and he also failed to sparkle at the fast Hockenheim circuit. But Merkel is right at home on the scratchers' circuits and come the final round at Manfeild, with a little bit of luck and some canny riding, he did everything required to secure the title.

The 26 year old Californian, who lives in Bergamo, Italy, can always be guaranteed to produce a fiery performance. He loves to entertain the crowd with his spectacular wheelies and has been a great ambassador for World Superbike racing in these formative years. Whether he turns to GP racing in 1990 or stays with Superbike, you just know Fred Merkel will continue his own success story.

# 2
# Stephane Mertens

Belgian Stephane Mertens must still be cursing his luck. Going into the final meeting at Manfeild, Mertens held a slender three point advantage but lost any chance of the title when a front disc broke up in the first race and pitched him off.

It was that sort of season, a season in which Mertens demonstrated his grit and determination to come back from serious injury only to end the year in bitter disappointment.

He missed the opening round at Donington after a serious crash at Brands Hatch on Good Friday. After rupturing the pelvic ligaments in a high speed crash with Joey Dunlop, doctors insisted he must rest but Mertens ignored their advice to take fifth and sixth places on the twisting Hungaroring circuit. Such was his pain, he had to be lifted on and off the machine for each race.

But Mertens stuck to his task and clawed his way back into contention for the title. Always in the top six in the next seven races, he scored a win in Austria then repeated in France and was suddenly the threat many experts had always predicted he could be.

The brake failure at Manfeild robbed him of the title and his 100% point scoring record in the ten rounds he contested, a fact which speaks highly, not only for his riding ability, but also that of his tuner Jean d'Hollander who fettled the machines.

Mertens, after a frustrating season with the unreliable Bimota, enjoyed a new lease of life in World Superbike after securing one of the three HRC supplied bikes in the series and safely laid to rest any reputation for crashing which he unfairly earned whilst with the Italian team the previous year.

# 3
# Raymond Roche

What a great season the former 500cc Grand Prix rider had on the fire engine red Ducati. The big vee twin was not always the epitome of reliability but when there were no problems with the electrics or the engine, the former Grand Prix star from Ollioules, France, was a consistent front runner.

On the fast tracks a combination of the Ducati top speed and his vast experience was unbeatable and he recorded double wins at Brainerd and Hockenheim, two of the fastest tracks on the World Superbike circuit.

The season didn't start too well for Roche. He retired in both races at the Donington opener then after finishing second at the switchback Hungaroring was again hit by electrical problems in the second leg.

But he soon put those disappointments behind him and came through to mount a serious challenge on the title. Poor wet weather performances at Sugo effectively ruined his chances even though he scored that memorable double win at Hockenheim the following meeting.

Roche won a total of five races, more than any other rider in the series to confirm that at 32, this ex-GP star still has plenty to offer in Superbike racing.

In 1990 Roche should consolidate his potential as a World Superbike Championship prospect by staying with the impressive Ducati team which will have further refined the twin cylinder machinery for the new campaign.

And by being joined by Italian Giancarlo Falappa Roche's chances of lifting the crown in 1990 must be vastly improved.

# 4
# Fabrizio Pirovano

Fabrizio Pirovano, the pint sized jockey on the Belgarda/ Byrd Yamaha OWO1, was not only one of the fastest starters in World Superbike, he also was a regular top six finisher - when he stayed aboard his motorcycle.

After his splendid runner up placing to Fred Merkel in the 1988 series, Pirovano's '89 fourth place finish was something of a disappointment and his season ended prematurely in Australia when he broke his ankle after a frightening crash with Malcolm Campbell.

Pirovano started the season well enough with a win in the first leg at Donington but slid off in the second and finished 18th after remounting. He never finished outside the top six in the next four meetings but can look back on the Japanese meeting at Sugo to find where his championship bid went awry.

He went into the race with a broken bone in his wrist from a crash at an Italian National Superbike round and then crashed in the first leg at Sugo and dnf'ed in the second.

Then he crashed again in the following meeting at Ricard. He was quickly back in the groove, again staying in the top six places until he crashed in the second leg at Oran Park, this time through no fault of his own.

This year Pirovano, a former motocrosser and the 1987 Italian Superbike Champion, struggled to find backing for his campaign after a major sponsorship deal fell through at the eleventh hour but managed to procure special factory parts for the Oran Park round as he still had an outside shot at the title. He badly needs factory backing and only then will he be able to show his true potential.

# 5
# Giancarlo Falappa

Mark this man down as one of the finds of the Superbike year. The Italian's hero, Giancarlo Falappa is without doubt the crowd's favourite. He even has his own following of 'Falappa Fans' who follow his exploits around the tracks of Europe. But this rider is not out to win just fans. He wants to win races!

Falappa is the wild man of Superbike. A real hard charger who stops at absolutely nothing in his quest to reach the chequered flag first.

His totally unorthodox style which sees him pitching the bike down into the turns, whilst remaining sat upright, makes him stand out in the pack.

Early last year, after his winning performance in the second leg at Donington, he was criticised for his aggressive riding but as the season progressed, Falappa smoothed off the rough edges and became an accomplished, but still very forceful performer.

He went on to win two more races and could have finished with a lot more points in the bag, had Bimota not opted out of the Australasian rounds.

Falappa only started road racing in 1987 after a career in motocross. In 1988 he lifted the Italian Sport Production title riding a Bimota YB4. For what the 25 year old lacked in experience last year he more than made in aggression and clearly has a great future in the sport.

# 6
# Anders Andersson

Swedish favourite Anders Andersson switched to Yamaha for the 1989 season, so ending a long association with Suzuki and experienced an immediate change in fortunes.

Quite simply, Andersson became one of the main threats in the World Superbike series after he had taken time to gain confidence with his new machinery.

Andersson's attention to detail on his OWO1 Yamaha is meticulous and his bike bristles with innovative ideas. He was one of the first to fit Ohlins inverted front forks which run in a very neatly machined set of machined triple clamps.

He also opted for the very powerful, Swedish made ISR brakes, front and rear and his engine is prepared in association with world famous drag racing specialist engineer, Thomas Klarkner.

Last year Andersson consistently finished among the top six, and was rewarded with fifth place in the final standings, but if there is any criticism of his riding it must be that he lacks that special 'fire' which will take him into the winners' circle.

Maybe after a year learning about the Yamaha, Andersson will go into the 1990 Superbike championship with a more aggressive approach. But for this seasoned veteran of international competition time for winning World Championships is running out.

# 7
# Terry Rymer

After his brilliant efforts in the 1988 World Superbike championship on a privately entered Honda, many expected young Englishman Terry Rymer to be the main threat to the established stars of the series.

Sure enough, the 22 year old Londoner top scored in the opening round at Donington Park, scoring third and second places on his Loctite Yamaha. But right at the end of the season after he faded from the leaderboard Rymer put in one of the sensational rides of the season to win the first race at Manfeild

Quite often he had been a front runner in the early going but his Dunlop rubber just didn't seem capable of lasting a Superbike length race but experimental tyres worked perfectly in New Zealand to end his season on a high note.

Rymer was also down on power to the similar Yamaha machines, the Loctite team having no time to spend on machine development during their season which was packed with domestic races as well as the Superbike rounds.

Even so he still took a good pair of fifth places at Brainerd, his first ever visit there, and then finished fourth at Paul Ricard in the opening race before crashing out of the second trying too hard on rubber which had gone off.

The team's heavy schedule meant Rymer missed the German and Sicilian races, a fact which didn't please him. He rejoined the fray in Australasia but crashed heavily on spilt oil at Oran Park injuring his hand. his performance in New Zealand, he also took fourth place in race two, was his saving grace.

Rymer has since re-signed for the Loctite team for a serious crack at the title in the forthcoming year. Given a fast bike, with good rubber, this young hot shot will more than live up to his potential in 1990.

# 8
# Baldassarre Monti

Ducati took a gamble when they signed 23 year old Baldassarre Monti at the start of the season, team boss Marco Lucchinelli admitting as much at Donington Park early in the season. But 1988 Italian Battle of the Twins champion surprised everyone in World Superbike with his talented riding aboard the potent vee twin and he finished a well deserved eighth in the final standings.

If anything, new boy Monti was held back by the Ducati camp. He surely could have improved on his eighth place in the standings if he had been allowed to do all the rounds. He missed the opener at Donington when the factory had no motorcycle built for him, then sat out the final two races as the team put everything into a final push for the title with Roche.

Like his Italian counterpart, Falappa, Monti is a relative newcomer to the sport. He only began racing in 1986 but has been willing to gain experience on any motorcycle. For instance in his BoTT winning year he also raced 125 Cagivas, RD350 Yamahas as well as the thundering Ducatis and won a total of 28 races in 37 starts.

If anything, he lacked consistency, finishing sometimes as high as third in one leg of a race and dropping out of the points in another. But with experience and the guidance of someone like Ducati team manger Marco Lucchinelli, he will come good. No wonder the Italians are already proclaiming his youngster 'Monti - the red terror'.

Clear evidence of Monti's potential came in the fifth round of the Series at the Osterreichring in Austria when he turned in a stirring performance to shatter team manager Marco Lucchinelli's existing lap record and further increase his credibility in the World Superbike Championship paddock.

# 9
# Jari Suhonen

Jari Suhonen, the ex-ice racer from Finland, made a slow start in the World Superbike championship last year when he raced his near standard OWO1 at Donington Park and finished out of the points in both races.

After that he worked hard to get his machine into competitive trim and it paid off with several point scoring rides, particularly in Pergusa where he finished sixth and fifth and eventually wound up tenth overall in the standings. A great effort from this essentially privateer entry.

His best ride, though the results won't show it, was in Austria where he was able to demonstrate his superb wet weather riding ability and ran with the front runners for many laps. He eventually finished tenth in the closely fought event. It was a great pity the Finn did not race in the wet Japanese round - another meeting where he could have put his special talents to good use.

If Andersson's Yamaha is the quickest of the OWO1s, then Jari's bike is not far behind. The Finn runs under the Arwidson banner, the same sponsors who once backed the late, great Jarno Saarinen.

Like Merkel, Suhonen runs a different brand of tyre to most. He opts for Metzeler rubber - getting special tyres from the German manufacturer - against the more popular choice of Michelin or, to a lesser degree, Dunlop.

Thanks to his sixth and seventh places in New Zealand, Suhonen moved to ninth overall in the championship, leap-frogging the absent Bimota rider, Mike Baldwin. With his bike sorted earlier this privateer could well improve on that in 1990.

# 10
# Michael Dowson

Australian Mike Dowson only contested three rounds of the championship but more than left his mark on the series. Riding in the Marlboro Yamaha Dealer Team colours of his home based sponsors, Dowson was given a factory prepared OWO1 that bristled with trick parts and was bang on the minimum weight limit.

He used it to the best possible advantage, placing second to Doug Polen in the opening Sugo race, then wining the second to become the top point scorer at the meeting even though a crash in practice left him with a very painful throttle hand.

In Australia on the same bike, he was seventh in the first leg but came through to win the second then got a fourth in the first Manfeild race and crashed in the second, through 'pilot error'.

Now 29, Dowson has twice been passed over whilst with MYDT, firstly at the end of 1987 when former teammate Kevin Magee was slotted into the Lucky Strike Yamaha GP squad and then last winter when Michael Doohan was plucked from the team and sat on a Rothmans Honda NSR500.

But Dowson has always been highly rated and last season alone also won the highly prestigious World Formula One race at Sugo in May. There is no doubt he is one of the top fourstroke racers in the world so it would be good to see him branching out into Europe for a full season on the Superbike circuit.

However, it seems likely that Dowson, like his former teammates, will be aiming for GP racing if he comes to Europe at all (he was considering an offer from Kawasaki to run a 250GP bike as we went to press) and that would be great loss for Superbike racing.

# Round 1

# Donington, Britain

## 27 March

Lucchinelli, in his new role as Ducati factory team manager, and skateboard driver.

The skies may have been blue but World Superbik

The French Kawasakis were neatly turned out bu

Swede Anders Andersson switched from Suzuki to Yamaha for the new season.

*tarted 1989 under a cloud following the withdrawal of the Sports Marketing Company.*

*acked speed at Donington.*

*Terry Rymer's Loctite Yamaha sported works kit swing arm and Dunlop rubber.*

Despite being amidst a cloud of political wrangles, the Donington Park opening round of the 1989 World Superbike Championship was a shining example of the exciting racing big four-stroke, silhouette machines can produce.

After qualifying, just 4.17 seconds split the forty rider line up with five of the six manufacturers represented in the series on the front two rows of the grid. Then in the first race, won by wily little Italian Fabrizio Pirovano, a blanket covered the first four finishers and little known Italian Giancarlo Falappa scored a shock victory in the second - his first ever race outside his home country.

Both races saw new lap records established, pole sitter Raymond Roche eventually getting his name in the record books in the second race with a 1m 13.26 on a Ducati twin which later expired with electrical problems as it had in the first.

Donington will also long be remembered by British fans as heralding the arrival of Britain's Terry Rymer as a world class competitor. He showed some of his potential in 1988 on a private Honda but came of age at Donington with a pair of classy rides on his factory Loctite Yamaha FZR750R OWO1 to take the equal point lead alongside his more experienced fellow countryman Roger Burnett

Donington showed the commitment to the series by importer teams and privateers alike with 55 entries accepted from 120 applications for the forty qualifying places. Honda, Yamaha, Suzuki and Kawasaki were all represented by importer teams with factory kitted machines while Ducati - with only a pair of 851S for Roche due to lack of time to build an extra bike their young starlet Baldasarre Monti - and Bimota both had full blown factory bikes

***ENTRY/QUALIFYING***

Roche sat on pole after the two qualifying sessions with an electric 1m 13.10 set late in the final session with the growling Ducati vee twin, the bright red machine now sporting inverted Ohlin forks, new beefed up swing arm, new style exhausts and more power from the 888cc motor.

The Frenchman's last minute effort bumped the leading Honda of Fred Merkel (1m 13.57) who had spent both sessions turning only a handful of laps at a time before frying his Pirelli qualifying rubber.

Burnett posted a 14.01 for fourth on the grid with his HRC bike, having spent qualifying and the Eurolantic challenge

*Falappa's unusual and forceful style earned him the 'frantic' tag at Donington.*

races, run throughout the Easter period, to set up the brand new bike with the help of suspension expert Ron Williams.

Between Merkel and Burnett on the grid was the unknown Falappa (1m 13.92), one of two new Bimota factory signings. The 25 year old Italian looked to have all the flair and aggression of his former peer Davide Tardozzi plus a rather unique, pitch it in but stay sitting upright, style.

Falappa, along with another ex-GP rider to join the Superbike ranks, Fabio Biliotti, had the latest update of the very rapid YB4EI, now with more profuse use of carbon fibre in order to keep the bike bang on the minimum FIM weight limit of 165Kg. Similar to the Ducati, it had Ohlin inverted forks and new triangulated bracing section on the swing arm.

Tardozzi, the former factory Bimota rider who blew his chances in the final round the previous year, when he crashed on the warm-up lap of the very last race, was still on a Bimota but as a privateer under the colours of Speed Shadows Racing.

He failed to start at Donington. Already nursing a broken collarbone injury sustained in a slow speed road bike tumble only 15 days previous, he took a spectacular highside exiting Goddard chicane and though not seriously hurt, opted out of the action. His best qualifying lap had been a 1m 15.42.

Final slot on the front row was taken by the lanky German Andreas Hofmann riding an '88 spec RC30 for Eckert Motorratechnik to a 1m 14.37. Little known in Superbike circles, he had given Roger Burnett a good run in the Assen World TT F1 race the previous year.

Next up were the two Minolta Suzuki France GSXR750Rs of World Endurance champions Thierry Crine (1m 14.38) and Herve Moineau (14.45) followed by Terry Rymer on the first of the OWO1s. Hampered by traffic and unable to get a clear lap, he posted a 14.48 with his Tony Scott tuned engine as opposed to the Yamaha prepped version. He was just 0.02 quicker than Swede Anders Andersson on the beautifully prepared Yamaha Sweden machine.

*Terry Rymer put his Donington track experience to good use and came away sharing the points lead with fellow Brit, Roger Burnett.*

While Rymer's Loctite Yamaha sported the factory kit parts like Andersson's bike, the Swedish machine also featured inverted Ohlin shocks mounted in some of the neatest machined triple clamps. Stopping power was aided by, made-in-Sweden ISR calipers.

Row three saw the Kawasaki France ZXR750s sandwiched between two English Hondas. Scot Steve Hislop was 10th fastest with a 14.53 despite losing a motor in one session and in a Eurolantic race on Sunday when he was run over by Rob McElnea's Honda.

Former 500GP racer Patrick Igoa clocked a 14.58, 0.10 quicker than his teammate Christophe Bouheben, both riders struggling to get to grips with the tardy, overweight ZXRs in their debut outing. World TTF1 champion Carl Fogarty sat the other side of the green 'not so mean' ZXRs with his Metzeler shod Appleby Glade Honda.

Row four saw Pirovano, handicapped through qualifying with various handling and carburation problems, he could only post 16th fastest time on the new FJ Valli/Belgarda Yamaha OWO1. Next to him sat British Superstock champion Brian Morrison who also had a frenetic time with a bunch of carburation problems.

Still further back was Aussie hard man Malcolm Campbell after a Donington debut he would rather forget. He had crashed twice in practice, both times at high speed and lined up back on the seventh row, 28th fastest, with a battered looking Citizen Honda and an equally battered body. His injuries included a badly swollen right hand which

*Roger Burnett leads Terry Rymer, with Giancarlo Falappa out of shape in the background and Fabrizio Pirovano running for cover.*

made throttle and braking control almost impossible and a badly bashed knee which restricted his movement on the bike.

***RACE 1***

With almost 20,000 fans basking in the unseasonally mild Easter weather, Merkel holeshot the first 30 lap race, leading Falappa into Redgate. His lead was shortlived as the Italian barged by on the brakes for Goddard at the end of lap one.

Roche was already touring, his vee twin now a single with a broken ignition. he complete a couple more laps then pulled into the pits.

Falappa and Merkel were quickly joined by Pirovano from the third row, after yet another of his famous fast starts, Burnett, Rymer and Crine. A few yards further back came Andersson, a brief gap then the bunch led by Hofmann and courageous Mark Phillips, the Loctite Yamaha rider who had almost lost his life at this very circuit the previous September and was still fighting to regain his full strength after a long recuperation period.

The front runners immediately settled into a no holds barred fight for the lead with fairing bashing antics, more akin to a Pro-Am sprint, the order of the day. On lap nine Pirovano squeezed by Falappa into the lead at Goddard but a couple laps later at the Old Hairpin, the two clashed allowing Burnett and Rymer to slip through.

By third distance Crine had faded, his Suzuki not up to the task and was soon joined by Merkel - his tyres already overworked by the fast pace.

Lap eleven saw Falappa try a wild inside move to pass Pirovano, again at Goddard, forcibly moving the Yamaha off line. Two laps later The Bimota rider outbraked Burnett to retake the lead at the same spot and almost highside it. But most of his wild looking antics were caused by petrol spilling onto the rear Michelin.

''There was petrol coming out of the breather, directly onto the rear tyre but every time he got sideways he just wound the throttle on more,'' exclaimed Burnett.

The no quarter battle raged until lap 22 when Falappa tried to follow Pirovano through a gap under a lapped rider, Ramio Vartiainen at Coppice. By the time his Bimota front wheel arrived at the corner apex there was no gap, the lapee already having turned in. Falappa committed to turning, hit him midships and sent him crashing while he went into the sandtrap. Although Falappa managed to stay on two wheels he could only resume in 19th place his chance of points, let alone winning, now gone.

That incident let Pirovano build an all important cushion of a few yards which he held to the finish while Burnett pulled away from Rymer's Yamaha towards the end as the OWO1's Dunlops began to squirm around.

Merkel had let Crine lead for much of the last third of the race but slipped by with a couple of laps left.

Andersson who had been running with Fogarty for most of the race got by on the run in to the line. Andersson admitted he had lost the tow from the leading bunch in the early stages while he made sure his tyres were up to working temperature. Meanwhile Fogarty had run the entire race without a steering damper, the mounting bracket having stripped it's threads on the warm up lap..

***RACE 2***

Merkel again smoked the field into turn one but at the end of the lap it was Crine leading for Suzuki - but only briefly as Pirovano came through to take up the running.

Roche this time was midfield and picking up places fast while Campbell toured into retire. He had a big momemt on the first lap and decided it wasn't worth risking another crash during the weekend.

Lap two brought dramas as McElnea went down exiting Goddard and Spaniard Pardo went down in the ensuing confusion.

With corner workers still cleaning up there, Crine got off the Suzuki in a big way at Coppice toward the end of the next lap which saw his bike cartwheeling down the straight among the pack of riders. According to the Frenchman, Rymer had run him off the track but the Loctite Yamaha rider saw it different.

''Crine was off line so I went under him,'' said Rymer. ''He drifted out wide then came back down on me and grabbed a big handful.''

The Suzuki still on the marbles bordering the racing line suddenly lost traction and spit the rider over the bars.

The crash split the pack leaving Pirovano and Rymer clear with Falappa chasing hard and looking much more composed this time out. He snatched the lead from Pirovano with Brits Burnett and Rymer again in close attendance. Burnett was also menacing as was Roche on the Ducati, swiftly making ground after a poor start.

Lap eight and Bouheben crashed at Goddard in the same spot as McElnea and sustained a broken wrist. Then Hofmann went down at McLeans and the resulting confusion served to break up the flow of the race before it had a chance to settle into a rythmn.

There was no luck either for Roche. Around the 18 lap mark his vee twin went sour, again with an ignition failure but only after the French ace had turned a 13.26 record lap.

Falappa was gradually edging away while the remaining trio of Rymer, Burnett and Pirovano slugged it out for second. With around 7 laps left Pirovano went clear and looked to be closing on the leader but another lap saw the trio together again.

With only two laps left Pirovano dumped it at the old Hairpin. He got off line at the Old Hairpin trying to outbrake Rymer and found some oil. The front end tucked under and he was down on his ear in the dirt.

That left Rymer in second and Burnett a tired third but nonetheless happy with his joint points lead. The Honda rider admitted, ''I had a minor gearbox problem plus I was tired after doing the three Eurolantic support races the same day so I decided to play safe for points rather than risk everything trying to pass Terry.''

Hislop came into fourth from Andersson with Merkel sixth just holding off the West German Suzuki rider Gschwender.

*Pirovano drew first blood in the World Superbike Championship but only after a hectic scrap in the opening race at Donington.*

*Raymond Roche was fast at Donington but the Ducati electrical system proved unreliable.*

DUCATI
Superbike
Agip
MICHELIN
WEBER
Superbike

# Round 1

# Donington, Britain

## 27 March

# Race Data

# Qualifiers

| POS | RIDER/NATIONALITY | ENTRANT/BIKE | TIME |
|---|---|---|---|
| 1. | Raymond Roche (F) | Squadra Corse Ducati 851S | 1m 13.10 |
| 2. | Fred Merkel (USA) | RCM/Rumi Honda RC30 | 1m 13.57 |
| 3. | Giancarlo Falappa (I) | Bimota Spa YB4EI | 1m 13.92 |
| 4. | Roger Burnett (GB) | HRC Honda RC30 | 1m 14.01 |
| 5. | Andreas Hofmann (D) | Eckert Honda | 1m 14.37 |
| 6. | Thierry Crine (F) | Minolta Suzuki GSXR750R | 1m 14.38 |
| 7. | Herve Moineau (F) | Minolta Suzuki GSXR750R | 1m 14.45 |
| 8. | Terry Rymer (GB) | Loctite Yamaha FZR750R | 1m 14.48 |
| 9. | Anders Andersson (S) | Yamaha Sweden FZR750R | 1m 14.50 |
| 10. | Steve Hislop (GB) | Honda UK RC30 | 1m 14.53 |
| 11. | Patrick Igoa (F) | Kawasaki France ZXR750 | 1m 14.58 |
| 12. | Christophe Bouheben (F) | Kawasaki France ZXR750 | 1m 14.68 |
| 13. | Carl Fogarty (GB) | Appleby Glade Honda RC30 | 1m 14.73 |
| 14. | Mark Phillips (GB) | Loctite Yamaha FZR750R | 1m 14.76 |
| 15. | Peter Rubatto (D) | Hein Gericke Bimota YB4EI | 1m 14.80 |
| 16. | Fabrizio Pirovano (I) | Belgarda Yamaha FZR750R | 1m 14.83 |
| 17. | Brian Morrison (GB) | Honda UK RC30 | 1m 14.89 |
| 18. | Ernst Gschwender (D) | Suzuki Deutschland GSXR750R | 1m 15.08 |
| 19. | Fabio Biliotti (I) | Bimota Spa YB4EI | 1m 15.20 |
| 20. | Rob McElnea (GB) | Padgetts Honda RC30 | 1m 15.49 |
| 21. | Andy McGladdery(GB) | Francis Neill Honda RC30 | 1m 15.57 |
| 22. | Paul Iddon (GB) | Pee Jay Yamaha FZR750R | 1m 15.60 |
| 23. | Rene Delaby (L) | Honda RC30 | 1m 15.61 |
| 24. | Tommy Douglas (CDN) | Team Arom Honda RC30 | 1m 15.63 |
| 25. | Jari Suhonen (SF) | Arwidson Yamaha FZR750R | 1m 15.67 |
| 26. | Rene Rasmussen (DK) | Pepsi Suzuki GSXR750R | 1m 15.73 |
| 27. | Robert Dunlop (IRL) | O'Kane Honda RC30 | 1m 15.76 |
| 28. | Malcolm Campbell (AUS) | Citizen Honda RC30 | 1m 16.10 |
| 29. | Peter Skold (S) | Honda RC30 | 1m 16.27 |
| 30. | Anton Gruschka (D) | Honda RC30 | 1m 16.39 |
| 31. | Mark Linscott (GB) | Francis Neill Honda RC30 | 1m 16.41 |
| 32. | Jeffrey de Vries (NL) | Yamaha FZR750R | 1m 16.43 |
| 33. | Michael Galinski (D) | Mitsui Yamaha FZR750R | 1m 16.45 |
| 34. | Marino Fabbri (I) | RB Moto Kawasaki ZXR750 | 1m 16.52 |
| 35. | Dave Leach (GB) | Yamaha FZR750R | 1m 16.71 |
| 36. | Paul Lewis (AUS) | Francis Neill Honda RC30 | 1m 16.83 |
| 37. | Johan v Vaerenbergh (B) | Kawasaki Belgium ZXR750 | 1m 16.92 |
| 38. | Jose M Pardo (E) | Honda RC30 | 1m 17.15 |
| 39. | Raimo Vartiainen (SF) | Suzuki GSXR750R | 1m 17.26 |
| 40. | Piergiorgio Bontempi (I) | RCM/Rumi Honda RC30 | 1m 17.27 |

| | |
|---|---|
| **Race Data:** | 2 x 30 laps of 1.957 mile circuit |
| **Total Distance:** | 117.42 miles |
| **Weather:** | Cool, Dry |
| **Fastest Lap:** | Raymond Roche (race two) 1m 13.26s, 96.16mph, 154.75kph |
| **Existing Lap Record:** | Bubba Shobert 1m 14.10s, 95.07mph, 152.99kph (April 1988) |

# Results Race 1

| POS | RIDER | LAPS | TIME | MPH | KPH |
|---|---|---|---|---|---|
| 1. | Pirovano | 30 | 37m 14.42 | 94.69 | 152.25 |
| 2. | Burnett | 30 | 37m 14.42 | 94.47 | 152.02 |
| 3. | Rymer | 30 | 37m 17.10 | 94.37 | 151.88 |
| 4. | Merkel | 30 | 37m 26.40 | 94.10 | 151.44 |
| 5. | Crine | 30 | 37m 27.04 | 94.07 | 151.39 |
| 6. | Andersson | 30 | 37m 28.44 | 94.01 | 151.30 |
| 7. | Fogarty | 30 | 37m 28.61 | 94.00 | 151.29 |
| 8. | Morrison | 30 | 37m 30.28 | 93.93 | 151.17 |
| 9. | Hofmann | 30 | 37m 30.44 | 93.93 | 151.16 |
| 10. | Igoa | 30 | 37m 38.37 | 93.60 | 150.63 |
| 11. | Gschwender | 30 | 37m 38.47 | 93.59 | 150.63 |
| 12. | Phillips | 30 | 37m 47.91 | 93.20 | 150.00 |
| 13. | Rubatto | 30 | 37m 50.15 | 93.11 | 149.85 |
| 14. | Moineau | 30 | 37m 50.90 | 93.08 | 149.80 |
| 15. | Bouheben | 30 | 37m 56.05 | 92.87 | 149.46 |
| 16. | McElnea | 30 | 37m 56.34 | 92.86 | 149.44 |
| 17. | Falappa | 30 | 38m 07.78 | 92.39 | 148.70 |
| 18. | Iddon | 30 | 38m 08.29 | 92.37 | 148.66 |
| 19. | Campbell | 30 | 38m 12.39 | 92.21 | 148.40 |
| 20. | McGladdery | 30 | 38m 13.59 | 92.16 | 148.32 |
| 21. | Delaby | 30 | 38m 13.72 | 92.15 | 148.31 |
| 22. | Galinski | 30 | 38m 21.75 | 91.83 | 147.79 |
| 23. | Suhonen | 30 | 38m 21.80 | 91.83 | 147.79 |
| 24. | de Vries | 30 | 38m 22.15 | 91.82 | 147.77 |
| 25. | Rasmussen | 30 | 38m 26.32 | 91.65 | 147.50 |
| 26. | Dunlop | 30 | 38m 26.39 | 91.65 | 147.50 |
| 27. | Pardo | 30 | 38m 26.67 | 91.64 | 147.48 |
| 28. | Gruschka | 30 | 38m 27.03 | 91.62 | 147.46 |
| 29. | Bontempi | 29 | | | |
| 30. | Biliotti | 29 | | | |
| 31. | Leach | 29 | | | |

**NOT CLASSIFIED**

Vartiainen 20 laps, Douglas 16, Fabbri 13, Hislop 11, Skold 11, Lewis 9, Vaerenbergh 4, Roche 3, Linscott 0

# Results Race 2

| POS | RIDER | LAPS | TIME | MPH | KPH |
|---|---|---|---|---|---|
| 1. | Falappa | 30 | 37m 16.10 | 94.51 | 152.09 |
| 2. | Rymer | 30 | 37m 16.80 | 94.49 | 152.04 |
| 3. | Burnett | 30 | 37m 24.91 | 94.14 | 151.49 |
| 4. | Hislop | 30 | 37m 31.28 | 93.88 | 151.06 |
| 5. | Andersson | 30 | 37m 33.77 | 93.77 | 150.89 |
| 6. | Merkel | 30 | 37m 49.56 | 93.12 | 149.84 |
| 7. | Gschwender | 30 | 37m 50.11 | 93.10 | 149.81 |
| 8. | Rubatto | 30 | 37m 56.56 | 92.84 | 149.38 |
| 9. | Phillips | 30 | 37m 58.38 | 92.76 | 149.26 |
| 10. | Igoa | 30 | 38m 01.31 | 92.64 | 149.07 |
| 11. | McGladdery | 30 | 38m 04.94 | 92.49 | 148.84 |
| 12. | Iddon | 30 | 38m 06.61 | 92.43 | 148.73 |
| 13. | Fogarty | 30 | 38m 09.09 | 92.33 | 148.57 |
| 14. | Rasmussen | 30 | 38m 12.46 | 92.19 | 148.35 |
| 15. | Delaby | 30 | 38m 13.52 | 92.15 | 148.28 |
| 16. | Dunlop | 30 | 38m 14.35 | 92.12 | 148.23 |
| 17. | de Vries | 29 | | | |
| 18. | Pirovano | 29 | | | |
| 19. | Suhonen | 29 | | | |
| 20. | Leach | 29 | | | |
| 21. | Gruschka | 29 | | | |
| 22. | Skold | 29 | | | |

**NOT CLASSIFIED**

Douglas 27 laps, Roche 21, Morrison 18, Biliotti 16, Bontempi 13, Moineau 12, Hofmann 8, Bouheben 7, Crine 4, Campbell 1, McElnea 1, Pardo 1

# Championship Positions

**1.** Burnett 32, Rymer 32, **3.**Merkel 23, **4.** Andersson 21, **5.** Falappa, Pirovano 20, **7.** Gschwender 14, **8.** Hislop 13, **9.** Igoa, Fogarty 12, **11.** Crine, Rubatto, Phillips 11, **14.** Morrison 8, **15.** Hofmann 7, **16.** McGladdery 5, **17.** Iddon 4, **18.** Moineau 2, Rasmussen 2, **20.** Delaby, Bouheben 1.

# Round 2

# Hungaroring, Hungary

30 April

**(Above)** *The triple flags of Fred Merkel's camp.* **(Far Left)** *Fabrizio Pirovano (2) and Rob McElnea were locked in combat in both races.*

*Ducati newboy Monti leads Roche and Morrison*

*Merkel salutes the crowd after his storming performance on the switchback Hungaroring.*

*Rob McElnea was in sensational form on the stock Padgetts OWO1.*

*Merkel leads the bunch through the chicane.*

Californian Fred Merkel proved all the sceptics wrong when he dominated both races of the Hungarian Superbike round with his Pirelli shod Honda. Many doubted the tyres would last the distance. Maybe that was why no one gave chase when Merkel shot off into the lead? Maybe the man was just hot on the day?

Merkel had won the opening leg at the same circuit the previous year - a circuit where a good start is all important such is the tight, tortuous track where overtaking is oh so difficult.

Merkel's win gave him a handy 11 point lead in the series over Fabrizio Pirovano who finished the day with a third and second place in the bag. But that didn't come easy as he was pressured in both races by Rob McElnea, the Briton enjoying a new lease of life on the Padgett Yamaha.

Joint point leaders from Donington, Roger Burnett and Terry Rymer, both had bad times. Burnett crashing in one and non starting in the second while Rymer found it difficult to get to grips with the track.

The race also gave Ducati new signing Baldasarre Monti an opportunity to show his potential. He was fifth in the second race behind Italy's other new hotshot Giancarlo Falappa.

***ENTRY/QUALIFYING***

With over 120 entries for the Hungarian round accepted, 76 riders actually fronted for qualifying. Due to the huge turnout, the sessions were split into two groups.

West German Ernst Gschwender was the fastest of the first even numbers session on his Suzuki Deutschland GSXR750R. His 1m 50.83, nearly a second quicker than that of nearest rival, joint series point leader, Terry Rymer on the Loctite Yamaha, prompted suggestions of the German running oversized carbs.

It was interesting that the German rider was one of the few in his session who failed to improve his time in the second qualifying period. In fact his second time, a 51.43, would only have netted him 15th fastest on the grid instead of a second row start.

When his compatriot, Peter Rubatto posted a 1m 50.40 with the Hein Gericke Bimota to become the fastest of the second even numbers session that really got the tongues of his rivals wagging yet there was no official intervention to verify the claims so the times stood.

Flyin' Fred Merkel grabbed the pole with a superb 1m 48.86, a full 1.41 seconds faster than second man, Frenchman and ex-GP star Raymond Roche on the Ducati. The big question mark over the Californian's time on the RCM/Oscar Rumi HRC RC30, bearing in mind his Donington difficulties, was whether his Pirelli race tyres were as much up to the task as the qualifiers?

*Former GP racer Patrick Igoa rode hard for little return in Hungary.*

Although Merkel had a 1.41 cushion over Roche, the spread over the first four rows (18 riders) was just 3.51 seconds and the entire 40 bike grid were separated by just 4.91 seconds!

Next up was Rubatto then came another ex-GP rider, Britain's Rob McElnea. He provided one of the big shocks of qualifying when he scorched the brand new, and quite stock, Padgett Yamaha OWO1 around in 1m 50.48. Typifying the enthusiasm of his sponsors Peter and Clive Padgett, the Batley bike shop owners, they had bought him two such machines after the disappointing performance of their RC30 at Donington.

McElnea had had difficulty in becoming accustomed to the fourstroke characteristics at Donington but was well and truly on the pace at the tricky Hungaroring in spite of riding with the handicap of a broken wrist - legacy of his Donington crash.

After Gschwender came Italian Fabrizio Pirovano 0.21 slower than McElnea's Yamaha but obviously plenty of power in hand on the FJ Valli/Belgarda entered machine.

Malcolm Campbell was enjoying better fortunes in Hungary than those which befell him in England despite giving a bunch of horsepower to the opposition. His HRC engine had been shipped to Japan for a full rebuild so he fronted with stock RC30 with a '88 spec kit parts fitted.

Luckily the Hungaroring is not a horsepower track and his skilful performance in qualifying saw him only the seventh rider to break into the 50s with a 50.71.

Ever cheerful Dane, Rene Rasmussen, really did have something to smile about in Hungary. He was a second row starter with his Pepsi backed Suzuki GSXR750R with a

SHELL-INTERAG KFT.
VIDEOTON
Robottechnika

*The twisting Hungaroring puts a premium on machine handling rather than all out horsepower.*

51.10 and completing the row was the young Ducati hotshot, Baldasarre Monti - also with an impressive 51.10.

So where were all the other factory bikes? Joint championship point leaders Roger Burnett and Terry Rymer were on the third and fourth row respectively. Burnett on the HRC supplied RC30 did a 51.23 while Rymer, a rookie at this demanding circuit, openly admitted he was finding it difficult to put in a hot lap with the Loctite Yamaha and wound up with a 51.51.

The other HRC bike belonged to the Total/ Bel Ray/Team Mediares set up for Belgian Stephane Mertens. The fourth ranked World Superbike rider had missed Donington through injury and was still suffering from his Brands injuries. He had to lifted on and off the bike but still posted a noteworthy 51.81 on a track which required much physical effort from the riders.

Anders Andersson was a third row starter on the Yamaha Sweden machine as was the official Bimota Experience YB4EI of Giancarlo Falappa. The other Bimotas were not on the pace - if that's the right way to describe two riders only a couple seconds adrift of the front row.

Speed Shadows entered Davide Tardozzi was on the sixth row while Fabio Biliotti on the second factory Bimota was still struggling to adapt to fourstroke machinery and consequently was on row seven.

The two French Kawasakis looked after by Georges Godier were mid field qualifiers. Patrick Igoa hit a best of 51.87 while Pierre Bolle turned a 52.11. Bolle was in for the luckless Christophe Bouheben. This talented rider had broken a wrist in the Donington crash but it was a far worse car crash in France which ruled him out of the Hungaroring action.

The big no-show of the weekend was the Minolta Suzuki team of Thierry Crine and Herve Moineau. After their promising efforts at Donington many expected Suzuki France to show more commitment to World Superbike.

***RACE 1***

Pirovano pulled off yet another of his now famous starts to lead the field in the early going. But it was Merkel, winner of the first leg at the same circuit last year, who looked in control of the situation and breezed by after a handful of laps into a lead which never threatened throughout the remainder of the 25 laps.

That left Pirovano the task of fending off Raymond Roche and the thundering Ducati, plus McElnea who was riding the wheels off his Yamaha and Stephane Mertens at his courageous best on the HRC machine.

At middle distance the sixth place tussle between Rymer and Burnett, the joint championship points leader remember, became all out war as the Honda man dived under Rymer's Yamaha.

Burnett had been forcing the pace after a bad start and when the two collided it was him who went down in a heap. Rymer managed to hold his Yamaha upright. However, he lost a lot of time in his fight to control the wayward machine allowing Anders Andersson to move up a place. Burnett remounted to finish 32nd.

With Merkel off in the distance, ultimately to a seven second win, attention focussed on the battle for second place which was being resolved as Roche used the torque of the Ducati out of the tight turns to pull clear of

the other three.

Then Pirovano did likewise, only for McElnea to stage a last gasp effort which took him to 0.51 of a second at the flag. It was an interesting comparison with flyweight Pirovano on his kitted and fully sorted machine against the burly Humbersider on a stock, new, untried bike!

Mertens faded and finished a further five seconds adrift and was almost caught by Andersson who closed to within 0.16 at the flag. After the race the gritty Belgian had to be lifted from his machine, such was his pain and exhaustion after the gruelling 62 mile sprint.

Making the start at the Hungaroring is vital as several riders found to their cost. Reputations meant little in Hungary as names like Falappa (9th), Igoa (13th), Campbell (15th), Tardozzi (20th) floundered among the also rans.

***RACE 2***

There was drama before the second race started. Burnett on his spare machine had a head gasket blow on the warm up lap and dejectedly wheeled the bike into pit lane.

Merkel was even more dominant in the second race, leading from the start and winning by over 12 seconds - and nary a hint of the tyre troubles many expected. Certainly Merkel was in a class of his own on this particular weekend and he also smashed Lucchinelli's old lap record by some eight seconds - admittedly the old chicane had been removed for this year's race.

One person the Californian didn't have to worry about this time was Roche. He threw the Ducati away in the fast turn just after the pit straight at the end of the first lap.

Tardozzi too crashed out on lap two and Steve Hislop joined him on the list of non finishers with a blown fork seal. The Scot had only finish 19th in the first race - a bitterly disappointing weekend for the Honda UK rider after his promising Donington race two, fourth place.

Gschwender was another one to crash out. He had been dicing with Rymer for several laps and collided with him at the hairpin but actually crashed out at the same spot on his own a couple of laps later on the fifth lap. Rymer's progress was hampered by a couple of lurid slides which told him to back off the gas and settle for points. He wound up ninth.

Again the big action was for second place with McElnea once again giving Pirovano a great fight. Clearly the Briton was delighting in having a bike which wasn't blown away on the straights and was giving it everything to stay with the Italian.

Their wheel to wheel dice ended with Pirovano just in front at the chequered flag but only by 0.33 of a second.

In the early laps Italian youngsters Falappa and Baldasarre Monti on the Ducati had run with the two Yamahas but they were left to sort out their own private hot headed squabbles.

Falappa edged clear leaving the Ducati to the clutches of the fast moving Mertens who just failed to pip Monti by 0.68 of a second on the line.

Behind them came Andersson and Rasmussen, an impressive ride on the Pepsi Suzuki.

*Fred Merkel's Pirelli tyres worked perfectly allowing him to win the first race by over seven seconds and even bigger margin in the second.*

# Qualifiers

# Round 2

# Hungaroring, Hungary

## 30 April

# Race Data

| POS | RIDER/NATIONALITY | ENTRANT/BIKE | TIME |
|---|---|---|---|
| 1. | Fred Merkel (USA) | RCM/Rumi Honda RC30 | 1m 48.86 |
| 2. | Raymond Roche (F) | Squadra Corse Ducati 851S | 1m 50.27 |
| 3. | Peter Rubatto (D) | Hein Gericke Bimota YB4EI | 1m 50.40 |
| 4. | Rob McElnea (GB) | Padgett Yamaha FZR750R | 1m 50.48 |
| 5. | Ernst Gschwender (D) | Suzuki Deutschland GSXR750R | 1m 50.48 |
| 6. | Fabrizio Pirovano (I) | Belgarda Yamaha FZR750R | 1m 50.69 |
| 7. | Malcolm Campbell (AUS) | Citizen Honda RC30 | 1m 50.71 |
| 8. | Rene Rassmussen (DK) | Pepsi Suzuki GSXR750R | 1m 51.10 |
| 9. | Baldassare Monti (I) | Squadra Corse Ducati 851S | 1m 51.10 |
| 10. | Brian Morrison (GB) | Honda UK RC30 | 1m 51.12 |
| 11. | Roger Burnett (GB) | HRC Honda RC30 | 1m 51.23 |
| 12. | Andreas Hofmann (D) | Eckert Honda RC30 | 1m 51.24 |
| 13. | Anders Andersson (S) | Yamaha Sweden FZR750R | 1m 51.32 |
| 14. | Giancarlo Falappa (I) | Bimota Spa YB4EI | 1m 51.37 |
| 15. | Jari Suhonen (SF) | Arwidson Yamaha FZR750R | 1m 51.49 |
| 16. | Terry Rymer (GB) | Loctite Yamaha FZR750R | 1m 51.51 |
| 17. | Stephane Mertens (B) | Team Mediares Honda RC30 | 1m 51.81 |
| 18. | Jean-Yves Mounier (F) | Yamaha FZR750R | 1m 51.82 |
| 19. | Steve Hislop (GB) | Honda UK RC30 | 1m 51.83 |
| 20. | Patrick Igoa (F) | Kawasaki France ZXR750 | 1m 51.87 |
| 21. | Pierre Bolle (F) | Kawasaki France ZXR750 | 1m 52.11 |
| 22. | Mark Phillips (GB) | Loctite Yamaha FZR750R | 1m 52.29 |
| 23. | Jeffrey de Vries (NL) | Yamaha FZR750R | 1m 52.52 |
| 24. | Toni Rohrer (CH) | Honda RC30 | 1m 52.55 |
| 25. | Fabio Biliotti (I) | Bimota Spa YB4EI | 1m 52.59 |
| 26. | Paul Iddon (GB) | PJ Racing Yamaha FZR750R | 1m 52.73 |
| 27. | Micheal Galinski (D) | Mitsui Yamaha FZR750R | 1m 52.94 |
| 28. | Davide Tardozzi (I) | Speed Shadows Bimota YB4EI | 1m 53.12 |
| 29. | Dave Leach (GB) | Yamaha FZR750R | 1m 53.24 |
| 30. | Robert Holden (NZ) | Francis Neill Honda RC30 | 1m 53.26 |
| 31. | Andy McGladdery (GB) | Francis Neill Honda RC30 | 1m 53.31 |
| 32. | Wolfgang v Muralt (CH) | Honda Schweiz RC30 | 1m 53.39 |
| 33. | Peter Granath (S) | Yamaha Sweden FZR750R | 1m 53.41 |
| 34. | Rene Delaby (L) | Delaby Honda RC30 | 1m 53.41 |
| 35. | Darren Dixon (GB) | Francis Neill Honda RC30 | 1m 53.63 |
| 36. | Peter Skold (S) | Honda RC30 | 1m 53.67 |
| 37. | Christer Lindholm (S) | Suzuki GSXR750R | 1m 53.67 |
| 38. | Anton Grushka (D) | Honda RC30 | 1m 53.76 |
| 39. | Sven Seidel (D) | Suzuki GSXR750R | 1m 53.71 |
| 40. | Edwin Weibel (CH) | Kawasaki ZXR750 | 1m 53.77 |

| | |
|---|---|
| **Race Data:** | 2 x 25 laps of 2.497 mile circuit |
| **Total Distance:** | 124.74 miles |
| **Weather:** | Cloudy/Sunny Periods |
| **Fastest Lap:** | Fred Merkel (race two) 1m 49.20s, 81.29mph, 130.81kph |
| **Existing Lap Record:** | Marco Lucchinelli 1m 57.28s, 76.51mph, 123.13kph (April 1988) |

# Results Race 1

| POS | RIDER | LAPS | TIME | MPH | KPH |
|---|---|---|---|---|---|
| 1. | Merkel | 25 | 46m 29.67 | 79.54 | 128.01 |
| 2. | Roche | 25 | 46m 36.32 | 79.35 | 127.70 |
| 3. | Pirovano | 25 | 46m 40.02 | 79.25 | 127.53 |
| 4. | McElnea | 25 | 46m 40.71 | 79.23 | 127.50 |
| 5. | Mertens | 25 | 46m 45.73 | 79.09 | 127.28 |
| 6. | Andersson | 25 | 46m 45.89 | 79.08 | 127.27 |
| 7. | Rymer | 25 | 46m 53.12 | 78.88 | 126.94 |
| 8. | Gschwender | 25 | 46m 57.71 | 78.75 | 126.74 |
| 9. | Falappa | 25 | 47m 02.96 | 78.66 | 126.50 |
| 10. | Morrison | 25 | 47m 04.78 | 78.55 | 126.42 |
| 11. | Hofmann | 25 | 47m 17.21 | 78.21 | 125.86 |
| 12. | Rasmussen | 25 | 47m 21.55 | 78.08 | 125.67 |
| 13. | Igoa | 25 | 47m 21.62 | 78.08 | 125.67 |
| 14. | Suhonen | 25 | 47m 22.96 | 78.05 | 125.61 |
| 15. | Campbell | 25 | 47m 23.10 | 78.05 | 125.60 |
| 16. | Monti | 25 | 47m 26.22 | 78.03 | 125.57 |
| 17. | Dixon | 25 | 47m 34.61 | 77.73 | 125.10 |
| 18. | Bolle | 25 | 47m 45.72 | 77.43 | 124.61 |
| 19. | Hislop | 25 | 47m 48.57 | 77.36 | 124.49 |
| 20. | Tardozzi | 25 | 47m 51.49 | 77.27 | 124.36 |
| 21. | Mounier | 25 | 47m 51.73 | 77.27 | 124.35 |
| 22. | Gruschka | 25 | 47m 51.76 | 77.27 | 124.35 |
| 23. | Weibel | 25 | 47m 53.62 | 77.22 | 124.27 |
| 24. | Holden | 25 | 47m 55.32 | 77.18 | 124.20 |
| 25. | Rubatto | 25 | 47m 57.55 | 77.11 | 124.10 |
| 26. | v Muralt | 25 | 47m 59.36 | 77.06 | 124.02 |
| 27. | Skold | 25 | 47m 37.35 | 76.15 | 122.55 |
| 28. | Lindholm | 24 | | | |
| 29. | Granath | 24 | | | |
| 30. | Leach | 24 | | | |
| 31. | Delaby | 21 | | | |

**NOT CLASSIFIED**

Burnett 18 laps, Galinski 16, Iddon 16, Seidel 13, Biliotti 5, Rohrer 4, McGladdery 3, de Vries 3, Phillips 2

# Results Race 2

| POS | RIDER | LAPS | TIME | MPH | KPH |
|---|---|---|---|---|---|
| 1. | Merkel | 25 | 44m 14.57 | 80.25 | 129.14 |
| 2. | Pirovano | 25 | 44m 27.12 | 79.87 | 128.54 |
| 3. | McElnea | 25 | 44m 27.45 | 79.86 | 128.52 |
| 4. | Falappa | 25 | 44m 33.27 | 79.69 | 128.24 |
| 5. | Monti | 25 | 44m 38.31 | 79.54 | 128.00 |
| 6. | Mertens | 25 | 44m 38.99 | 79.51 | 127.97 |
| 7. | Andersson | 25 | 44m 39.13 | 79.51 | 127.96 |
| 8. | Rasmussen | 25 | 44m 40.16 | 79.48 | 127.91 |
| 9. | Rymer | 25 | 44m 44.46 | 79.36 | 127.71 |
| 10. | Rubatto | 25 | 44m 49.72 | 79.20 | 127.46 |
| 11. | Suhonen | 25 | 44m 54.15 | 79.07 | 127.25 |
| 12. | Campbell | 25 | 44m 56.93 | 78.98 | 127.11 |
| 13. | Morrison | 25 | 44m 59.72 | 78.90 | 126.98 |
| 14. | Hofmann | 25 | 45m 06.82 | 78.69 | 126.65 |
| 15. | Igoa | 25 | 45m 09.62 | 78.62 | 126.52 |
| 16. | Bolle | 25 | 45m 21.20 | 78.28 | 125.98 |
| 17. | Mounier | 25 | 45m 22.18 | 78.26 | 125.94 |
| 18. | Dixon | 25 | 45m 25.60 | 78.16 | 125.78 |
| 19. | Holden | 25 | 45m 27.94 | 78.09 | 125.67 |
| 20. | McGladdery | 25 | 45m 28.50 | 78.07 | 125.64 |
| 21. | de Vries | 25 | 45m 28.56 | 78.07 | 125.64 |
| 22. | Delaby | 25 | 45m 28.93 | 78.06 | 125.62 |
| 23. | Leach | 25 | 45m 29.97 | 78.03 | 125.58 |
| 24. | Rohrer | 25 | 45m 32.54 | 77.96 | 125.46 |
| 25. | v Muralt | 25 | 45m 37.53 | 77.81 | 125.23 |
| 26. | Biliotti | 25 | 45m 42.42 | 77.68 | 125.01 |
| 27. | Granath | 25 | 45m 45.60 | 77.58 | 124.86 |
| 28. | Weibel | 25 | 45m 58.79 | 77.21 | 124.26 |
| 29. | Skold | 24 | 45m 06.60 | 76.99 | 123.91 |
| 30. | Lindholm | 24 | 45m 41.21 | 76.14 | 122.53 |
| 31. | Phillips | 23 | | | |

**NOT CLASSIFIED**

Gruschka 11 laps, Gschwender 7, Seidel 4, Tardozzi 2, Hislop 2, Roche 1

# Championship Positions

**1.** Merkel 63, **2.** Pirovano 52, **3.** Rymer 48, **4.** Falappa, Andersson 40, **6.** Burnett 32, **7.** McElnea 28, **8.** Gschwender 22, **9.** Mertens 21, **10.** Morrison, Roche, Rubatto 17, **13.** Igoa 16, **14.** Rasmussen, Hofmann 14, **16.** Hislop 13, **17.** Fogarty 12, **18.** Crine, Phillips, Monti 11, **21.** Suhonen 7, **22.** Campbell, McGladdery 5, **24.** Iddon 4, **25.** Moineau 2, **26.** Delaby, Bouheben 1.

# Round 3

## Mosport, Canada

4 June

he journey
is the goal.
MOTOR
64
2
19

SHOEI
MICHELIN
64

GM

Californian-turned-Italian, Fred Merkel made believers out of the estimated 30,000 spectators at the inaugural North American round of the World Superbike Championship held at Canada's Mosport Park.

The 1988 World Superbike Champion rode his team Rumi Honda RC30 to the win in the first leg and a third-place finish in the second, racing both times with Canadian Superbike Champion Rueben McMurter, also on a Honda RC30, and in the second leg with Bimota-mounted Giancarlo Falappa as well.

Merkel led all but one lap of the first leg when McMurter got past briefly, then spent the second leg chasing Falappa, pulling alongside and even leading briefly, only to be denied by the aggressive Italian time and time again. Then a last-lap miscue cost Merkel second place when McMurter snuck by in the final turn.

After three rounds Merkel now leads the series with 98 points, followed by Italian Fabrizio Pirovano with 75, Falappa at 60, Britain's Terry Rymer 56 and Swede Anders Andersson with 52.

The race was held at the rider-proclaimed dangerous Mosport Park circuit which is located north of Lake Ontario in Canada's cottage country.

The scenic 2.459 mile, 10 turn circuit is laid out over rolling green hills and passes through a thick forest and is lined with unyielding Armco barriers and concrete walls with little or no run-off areas.

The general consensus among the visiting European riders was that there would have to be numerous changes made to the facility if they were to return in 1990.

The event marked the return of former multi-AMA Formula One Champion and 500cc grand prix competitor, Mike Baldwin, who finished eighth and sixth in the two legs. He would also be riding at the Brainerd, Minnesota round of the series for the Bimota Experience concern.

The question of whether the American superbikes, normally raced in the AMA Nationals, would be competitive with the World Championship Series superbikes (with their differing rules) was not really answered here as few Americans competed.

The best result, aside from the returning Baldwin, was achieved by Ohio's Tom Kipp who posted 16th and 18th finishes on his ageing, overweight (by World Superbike standards) Yamaha FZR750.

With over 20 Americans entered for the Brainerd round of the series perhaps the question would be answered there!

***ENTRY/QUALIFYING***

Qualifying, held under clearing skies (as were the races) saw Merkel post the quickest time (I minute 26.120 seconds) in the second of the Saturday afternoon sessions, cutting nearly a second off his best first-session effort (1:27.113) to take pole position.

Falappa was second best (1:26.189) while Mertens was third via his fastest first-session time (1:26.05). McMurter, the previous track record holder (1:28.55), qualified fourth (1m 26.834). The rest of the field turned 1.27s-plus with the 36th and final qualifying position going to Texan Jeff Harder.

Practice crashers on Friday included Baldwin who rode off on Moss' Corner (so-named after car racing legend Stirling Moss) the tight right hander that leads onto the Mario Andretti Straight. Baldwin attempted to steer back onto the course but toppled over unhurt when he clipped the hay bales.

"I went out on cold tyres and got a little too hot, "Baldwin admitted. "I'm not used to the radial front tyre, I've never used one before.

By the end of the weekend however the Connecticut resident had grown to like the radial. "I like the way the radial works, "he said, "A bias-ply tyre is faster into the corners but the radial is better coming out. You can steer the bike in at the exit."

Another Friday faller was Denmark's Rene Rasmussen who dropped his Kevin Schwantz-clone like number 34 Pepsi Suzuki GSXR750RK at the entrance to Moss' but fortunately escaped, aggravating his already broken collarbone.

*Mike Baldwin, seen here leading Canadian domicile*

Squadra Corse Ducati's newest recruit, Italian Baldassarre Monti joined the ranks of the fallen when he crashed in Saturday morning's practice prior to the first qualifying session; the only damage was to the bike's footrest.

Canadian Jacques Guenette Jr was the

*'iwi Gary Goodfellow, quickly got into the Superbike groove on the factory Bimota.*

only rider to suffer an injury when his Kawasaki ZX7's motor tied up and spit him off during qualifying. He was later seen sporting a sling on his right arm and would sit out the races.

***RACE 1***

Merkel began his conquest in fine form by taking the lead at the start of the first leg. Miguel DuHamel followed the champ closely on his Suzuki but fell on the second lap when he hit a bump while trying to pass Merkel around the outside of corner three.

''I thought the pace was too slow , so I tried to pass him, ''said the youthful French Canadian matter-of-factly. ''But there's a bump there that I didn't see. I'm a road racer on a motocross track!''

DuHamel's Suzuki carburettors inhaled dirt, so he opted to ride his production-class Suzuki in the second race. He did replace the essentially stock machine's road tyres with racing slicks however.

McMurter chased Merkel, followed, albeit some ways back by Pirovano, Mertens was next, with Andersson, Roche and Rymer all in hot pursuit.

The lead duo left the field behind and on lap seven McMurter moved into the lead, Merkel wouldn't allow it, however, and repassed McMurter before the completion of the lap.

Roche, victim of a bad start, wasted no time in moving up to third place, while Canadian Steve Crevier was similarly advancing man-by-man and soon found himself in the top 10.

*Home favourite Rueben McMurter relished the challenge of the barrier lined Mosport track.*

Rymer battled on his underpowered Loctite Yamaha OWO1 but his race would end on lap 17 with a spectacular crash in the fast Armco-lined, downhill turn one, probably the most dangerous portion of the track.

I was riding rings around everyone but getting passed by three or four riders at a time on the straight, ''Rymer complained later, limping but otherwise unhurt. ''The front end gave way; I was getting wider and wider (in turn one), I slipped and clipped a hay bale. It's so upsetting.''

Rymer's bike suffered a damaged footrest and lower fairing but he'd make repairs before the second race.

Falappa, meanwhile, passed Pirovano for the fourth position and was exchanging third place with Roche when he retired with a limp wrist caused by an arm protector which was too tight under his leathers. Falappa was joined in the pits by team mate Fabio Biliotti, who suffered from an intermittent ignition.

At this point Merkel had stretched out a six-second advantage which he would hold to the chequered flag. Roche began closing on McMurter, while Mertens ran fourth followed by Pirovano, who was beginning to feel pressure from the advancing Crevier.

Lap 20 saw Roche's inevitable pass over McMurter for the runner-up position, the Frenchman using the Ducati's abundant horsepower to pass the Canadian on the long uphill back straight.

''I was sliding really bad off turn three, I couldn't get on the gas early, '' said McMurter. ''That's one of the places I was making time.''

''I had the best machine on the race track, '' said Roche in broken English after the race. He would not be so lucky in the second leg.

On the 25th and final lap, Crevier drove up the inside of Pirovano at the exit of turn three to take fifth place, but the diminutive Italian motored by the Canadian on the back straight. Undaunted, Crevier used an inside line in the final turn to take fifth position.

''I was down on power so I had to get him in the turns, '' Crevier said later, ''I went right under him. It must have pissed him off!''

At the finish it was Merkel posting a comfortable win with Roche holding second just ahead of McMurter. Mertens claimed a lonely fourth, with the Yamaha OWO1-riding duo of Crevier and Pirovano fifth and sixth.

Finland's Jari Suhonen (Arwidsson Yamaha) edged Baldwin for seventh, with Andersson taking ninth and Crevier's Team Ciampini team mate, Tommy Douglas completing the top 10.

Canada's Michel Mercier (Suzuki), Britain's Paul Iddon (Yamaha), Rasmussen, Switzerland's Wolfgang von Muralt (Honda-Suisse) and Italy's Marino Fabbri (Bimota) rounded out the 15 FIM points-scoring positions.

***RACE 2***

The second race was even more action-packed than the first with drama unfolding even before the start. McMurter apparently stalled out on the course during the warm-up lap, necessitating an additional countdown and causing the race to be shortened to 24 laps due to fuel considerations.

Douglas was also stranded out on the course with an inoperative fuel pump but unlike McMurter he would not be able to make the start.

Merkel again led at the green light but this time was chased relentlessly by Falappa and McMurter. Pirovano held fourth followed by Suhonen and Monti, who missed the first race when his Ducati suffered yet another broken electrical connector during the sighting lap.

Team mate Roche's race was essentially over before it started, his Ducati suffering a

*Neither Briton, Terry Rymer (10) or Jari Suhonen (19) from Finland had a memorable time in Canada.*

broken cambelt tensioner bearing. He retired on lap seven.

Falappa, now with the elbow padding removed from his leathers , assumed the lead on lap three and held it until Merkel got by on the inside of the high-speed turn eight on lap 18.

Falappa reclaimed the lead on lap 21, using the same move that Flyin' Fred had used on him, after which Merkel tried each and every lap to gain an advantage at the end of the back straight. Falappa, however, was superior going into the following turn eight and continued to hold the lead.

The battle went this was until the final lap when Merkel's last- ditch effort was thwarted by a lapped rider. Merkel tried to go under Falappa at the exit of the final turn, but in the end it cost him a position rather than gaining him one.

''I get paid to stick in there, '' Merkel said afterwards. ''I knew I'd get either first or third. We were handlebar to handlebar: I went a little wide and Rueben got under me.''

''Fred bobbled,'' said McMurter. ''We came pretty close together but I had the advantage. I was hanging back hoping something would happen my way; something did!''

Farther back a titanic battle for fifth through 11th positions involving Suhonen, Monti, Mertens, Canadian Gary Goodfellow, Rymer, Crevier and Baldwin. After a heated on-track debate, it was Mertens, using a brilliant inside pass on Baldwin in turn eight on the final lap, who claimed fifth place.

''He was a little too wide,'' said Mertens of Baldwin. ''I thought also I pass Pirovano, but I thought better to finish fifth then no finish.''

Baldwin inched Monti for sixth, with Rymer following in eighth, Suhonen ninth and Goodfellow 10th. Goodfellow, incidentally, was riding with injuries received back in March at Daytona and more recently from the previous weekend's Canadian National at Sanair. His Suzuki had retired from the first race with a dropped valve.

Andersson and Biliotti took 11th and 12th, while Tardozzi, who'd retired from the first race with unknown electrical problems, finish 13th. Mercier and Rasmussen rounded out the top 15

For Baldwin, the event was a satisfying one, although he was really looking forward to the following week's Brainerd race: ''It's good to be back ,'' he said after the racing was done. ''I couldn't pass anyone in the first race , so we lowered the gearing for the second race and it was much better. Some days you've got to be the guy that gets passed; it was my turn today.''

After posting a fine fifth-place finish in the first race, Crevier led Goodfellow in the second until his motor began smoking and slowing down. He retired on lap 21.

Britain's Paul Iddon also retired after an insect (a ''Mosport Puckerbug'') splattered on his helmet's faceshield, impairing his vision.

Compatriot Roger Burnett fared no better retiring from both races when he realised that he was riding as hard as possible and getting nowhere. ''A terrible day'', said Burnett. In fairness it should be noted that Burnett was using his back-up motor after his race motor blew up in practice.

''This was my first time in Canada and I had a wonderful time,'' said Merkel as he stood atop the rostrum under the Italian flag, after the playing of the Italian National Anthem and he praised his rival McMurter.

''I've got to hand it to Rueben, he rode fantastic. I've ridden against him a bunch of times and he never rode like this. The race was obviously one of McMurter's proudest moments. I could hear you guys cheering,'' he told the fans.

Mertens, when asked why he couldn't catch the leaders , was humbled. ''I tried to but Fred and...what's his name...Rueben...were too fast for me.''

Obviously McMurter had made an impression on the Belgian, who vowed to remember the Canadian's name next time

*Defending champion Fred Merkel moved into a 23*

they met.

Falappa tried to explain what happened in the first race (he speaks very little English) but when he pointed to his elbow Merkel, who was acting as translator, pointed to the Italian's crotch and said, ''Oh, I thought it (the problem) was down there!'' Much laughter ensued at the expense of an embarrassed Falappa.

Following the victory celebration, CMA officials randomly selected McMurter's machine for a teardown-checking bore and stroke, carburettor size, as well as sound testing-and the machine was found to be legal.

*point lead in the championship after Mosport.*

# Qualifiers

# Round 3

# Mosport, Canada

# 4 June

# Race Data

| POS | RIDER/NATIONALITY | ENTRANT/BIKE | TIME |
|---|---|---|---|
| 1. | Fred Merkel (USA) | RCM/Rumi Honda RC30 | 1m 26.12 |
| 2. | Giancarlo Falappa (I) | Bimota Spa YB4EI | 1m 26.18 |
| 3. | Stephane Mertens (B) | Team Mediares Honda RC30 | 1m 26.50 |
| 4. | Rueben McMurter (CDN) | Action Accessories Honda | 1m 26.83 |
| 5. | Jari Suhonen (SF) | Arwidson Yamaha FZR750R | 1m 27.13 |
| 6. | Marino Fabbri (I) | Bimota | 1m 27.22 |
| 7. | Miguel Duhamel (CDN) | Suzuki GSXR750R | 1m 27.31 |
| 8. | Baldassarre Monti (I) | Squadra Corse Ducati 851S | 1m 27.36 |
| 9. | Raymond Roche (F) | Squadra Corse Ducati 851S | 1m 27.41 |
| 10. | Gary Goodfellow (CDN) | Don Knit/Yoshimura Suzuki | 1m 27.51 |
| 11. | Terry Rymer (GB) | Loctite Yamaha FZR750R | 1m 27.51 |
| 12. | Anders Andersson (S) | Yamaha Sweden FZR750R | 1m 27.57 |
| 13. | Fabrizio Pirovano (I) | Belgarda Yamaha FZR750R | 1m 27.62 |
| 14. | Roger Burnett (GB) | HRC Honda RC30 | 1m 27.94 |
| 15. | Davide Tardozzi (I) | Speed Shadows/Bimota | 1m 28.06 |
| 16. | Tommy Douglas (CDN) | Yamaha Canada FZR750R | 1m 28.18 |
| 17. | Mike Baldwin (USA) | Bimota Spa YB4EI | 1m 28.21 |
| 18. | Steve Crevier (CDN) | Yamaha FZR750R | 1m 28.33 |
| 19. | Michel Mercier (CDN) | Suzuki Canada GSXR750R | 1m 28.36 |
| 20. | Paul Iddon (GB) | Pee Jay Yamaha FZR750R | 1m 28.60 |
| 21. | Fabio Biliotti (I) | Bimota | 1m 28.65 |
| 22. | Rene Rasmussen (DK) | Pepsi Suzuki GSXR750R | 1m 28.93 |
| 23. | Wolfgang von Muralt (CH) | Honda Schweiz RC30 | 1m 29.02 |
| 24. | Tom Kipp (USA) | Yamaha FZR750R | 1m 29.30 |
| 25. | Rene Delaby (L) | Honda RC30 | 1m 29.71 |
| 26. | Steve Gervais (CDN) | Suzuki GSXR750R | 1m 30.83 |
| 27. | Yves Brisson (CDN) | Honda RC30 | 1m 31.62 |
| 28. | Derrick Medaglia (CDN) | Honda RC30 | 1m 32.54 |
| 29. | David Kieffer (USA) | Honda RC30 | 1m 32.68 |
| 30. | Steve Dick (CDN) | Suzuki GSXR750R | 1m 33.16 |
| 31. | James Adamo (USA) | Ducati | 1m 33.22 |
| 32. | Clyde MacDonald (CDN) | Kawasaki ZXR750R | 1m 33.81 |
| 33. | Don Vance (CDN) | Suzuki GSXR750R | 1m 33.87 |
| 34. | Johan v Vaerenbergh (B) | Kawasaki Belgium ZXR750R | 1m 34.70 |
| 35. | Jacques Guenette Jr (CDN) | Kawasaki ZXR750R | 1m 35.02 |
| 36. | Jeff Harder (USA) | Suzuki GSXR750R | 1m 36.85 |
| 37. | Oldrich Schmuttermeier (CDN) | Suzuki GSXR750R | 1m 37.45 |

**Race Data:** 2 x 25 laps of 2.459 mile circuit
**Total Distance:** 122.95 miles
**Weather:** Sunny, dry
**Fastest Lap:** Fred Merkel (race two) 1m 26.408s, 102.37 mph, 164.86 kph
**Existing Lap Record:** Establishes new class record

# Results Race 1

| POS | RIDER | LAPS | TIME | MPH | KPH |
|---|---|---|---|---|---|
| 1. | Merkel | 25 | 36m 35.51 | | |
| 2. | Roche | 25 | 36m 41.62 | | |
| 3. | McMurter | 25 | 36m 44.47 | | |
| 4. | Mertens | 25 | 36m 48.48 | | |
| 5. | Crevier | 25 | 36m 53.14 | | |
| 6. | Pirovano | 25 | 36m 53.31 | | |
| 7. | Suhonen | 25 | 36m 57.04 | | |
| 8. | Baldwin | 25 | 36m 57.21 | | |
| 9. | Andersson | 25 | 37m 04.64 | | |
| 10. | Douglas | 25 | 37m 14.95 | | |
| 11. | Mercier | 25 | 37m 37.58 | | |
| 12. | Iddon | 25 | 37m 38.19 | | |
| 13. | Rasmussen | 25 | 37m 42.91 | | |
| 14. | von Muralt | 25 | 37m 44.30 | | |
| 15. | Fabbri | 25 | 37m 44.71 | | |
| 16. | Kipp | 25 | 37m 48.53 | | |
| 17. | Brisson | 25 | 38m 06.67 | | |
| 18. | McDonald | 24 | 37m 35.65 | | |
| 19. | Medaglia | 24 | 37m 38.33 | | |
| 20. | van Vaerenberg | 24 | 37m 57.46 | | |
| 21. | Adamo | 24 | 38m 13.23 | | |
| 22. | Vance | 23 | 37m 19.33 | | |
| 23. | Dick | 21 | 37m 17.66 | | |

**NOT CLASSIFIED**

Harder 22, Gervais 20, Rymer 17, Falappa 17, Biliotti 15, Kieffer 14, Goodfellow 9, Burnett 6, Tardozzi 6, Delaby 5, Duhamel 1, Guenette Jnr 0, Monti 0

# Results Race 2

| POS | RIDER | LAPS | TIME | MPH | KPH |
|---|---|---|---|---|---|
| 1. | Falappa | 24.* | 34m 51.72 | | |
| 2. | McMurter | 24 | 34m 52.25 | | |
| 3. | Merkel | 24 | 34m 53.31 | | |
| 4. | Pirovano | 24 | 35m 05.03 | | |
| 5. | Mertens | 24 | 35m 05.22 | | |
| 6. | Baldwin | 24 | 35m 05.30 | | |
| 7. | Monti | 24 | 35m 05.70 | | |
| 8. | Rymer | 24 | 35m 10.78 | | |
| 9. | Suhonen | 24 | 35m 14.05 | | |
| 10. | Goodfellow | 24 | 35m 17.92 | | |
| 11. | Andersson | 24 | 35m 28.34 | | |
| 12. | Biliotti | 24 | 35m 38.41 | | |
| 13. | Tardozzi | 24 | 35m 40.53 | | |
| 14. | Mercier | 24 | 35m 40.68 | | |
| 15. | Rasmussen | 24 | 35m 40.92 | | |
| 16. | Duhamel | 24 | 35m 41.59 | | |
| 17. | Fabbri | 24 | 35m 50.42 | | |
| 18. | Kipp | 24 | 36m 50.39 | | |
| 19. | von Muralt | 24 | 36m 09.42 | | |
| 20. | Brisson | 24 | 36m 15.77 | | |
| 21. | Dick | 23 | 35m 53.68 | | |
| 22. | Gervais | 23 | 36m 01.34 | | |
| 23. | McDonald | 23 | 36m 02.08 | | |
| 24. | Medaglia | 23 | 36m 02.28 | | |
| 25. | Vance | 21 | 34m 54.01 | | |

* Race shortened by organisers due to warm-up lap problem.

**NOT CLASSIFIED**

Crevier 21 laps, van Vaerenberg 17, Iddon 8, Roche 7, Burnett 4, Adamo 3, Delaby 3, Kieffer 0, Harder 0, Douglas 0

*No further information supplied by race organisers*

# Championship Positions

**1.** Merkel 98, **2.** Pirovano 75, **3.** Falappa 60, **4.** Rymer 56, **5.** Andersson 52, **6.** Mertens 45, **7.** Roche 34, **8.** Burnett & McMurter 32, **10.** McElnea 28, **11.** Suhonen 23, **12.** Gschwender 22, **13.** Monti 20, **14.** Rasmussen, Baldwin 18, **16.** Morrisson, Rubatto 17, **18.** Igoa 16, **19.** Hofmann 14, **20.** Hislop 13, **21**. Fogarty 12, **22**. Crine, Phillips, Crevier 11, **25.** Iddon 8, **26.** Mercier 7, **27.** Goodfellow, Douglas 6, **29**. Campbell, McGladdery 5, **31**. Biliotti 4, **32.** Tardozzi 3, **33.** von Muralt, Moineau 2, **35.** Delaby, Bouheben, Fabbri 1,

# Round 4

# Brainerd, United States

## 11 June

RIVERVIEW
MOTEL
WELCOME
RACE FANS
AAA
APPROVED

HONDA
HONDA

Ducati-mounted Frenchman Raymond Roche beat the American contingent, including defending World Superbike Champion Fred Merkel, to score an upset victory at the inaugural U.S. round of the series at Brainerd International Raceway.

The highly-partisan race-day crowd of approximately 20,000 was hoping for an American sweep (in best Laguna Seca tradition), but instead the fans were treated to a French lesson as Roche scored his first ever wins in World Superbike competition, one in each of the two 20-lap races.

Leg one saw the 32-year-old ex-Grand Prix star from Le Beausset, France work up from a mediocre start, disposing of Honda RC30 riders Merkel and Belgian Stephane Mertens and Yamaha OWO1 mounted Italian Fabrizio Pirovano, en route to victory. Leg two saw Roche lead laps two to five, then make a mistake and drop to sixth before again working his way past Merkel and Mertens for the win.

Mertens took second overall via his 3-2 finishing position, while Pirovano took third overall with a 2-4 tally, Merkel scored a fourth in the first leg and third in the second for fourth overall on the day and continued to top the standings with 126 points.

The biggest disappointment for U.S. fans was the lack of competitive American riders. Those expecting Yoshimura's young lions Scott Russell and Jamie James or Vance and Hines David Sadowski to provide the fireworks were in for a shock as none of the three had their regular rides.

The reasons given for the major American teams' absences revolved around the differences between the AMA and FIM Superbike rules. The AMA permits one-millimetre over-bores while the FIM requires 750cc maximum displacement, and the FIM also requires the use of stock carburettors. In essence, a team would have to build a new motor for the race, something very few teams were willing to do.

So Russell entered his Dunlop DOT tyre-shod 88 Supersport bike, posting 15-17 finishes while Sadowski completed one

***(Left)*** *Roche leads Rymer, Mertens, Merkel, Pirovano and Monti.* ***(Below)*** *Falappa bites the dust.*

lap of leg one and finished just ahead of Russell in leg two on his privateer Suzuki. James tried to hustle up a ride, but in the end would sit out the Superbike races. Hypercycles Carry Andrew finished right behind Russell in both legs on his privateer Suzuki.

Aside from Merkel and Bimota's Mike Baldwin (7-23), the best performance by an American national rider was turned in by California's Richard Arnaiz on the Team Motorsport Yamaha, who finished in 10th in leg one before retiring from leg two with an overheating motor.

Unfortunately Arnaiz' machine was selected for a random teardown, and since he was running AMA-spec cylinders and Yamaha kit carbs, Arnaiz opted not to be torndown. He was disqualified from the event, fined $250 and suspended from FIM competition for a period of 30 days.

The next best performance turned in by an American was Doug Chandler on the Rob Muzzy-tuned Kawasaki ZX7, who sat out the first leg in order to save himself for the 750cc Supersport National (held between the two legs of World Superbike competition), and then finished 12th in the second leg. Ironically, Chandler was also subject to a teardown after a protest was filed by Italian Marino Fabbri, who had finished two positions behind Chandler in leg two and was looking to gain some world championship points.

Muzzy counter-protested Fabbri's Bimota saying, "If I've got to tear mine down, so will he." The bore and stroke of Chandler's Kawasaki and Fabbri's Bimota were checked and both bikes were found to be legal.

Best US finisher at Mosport, Canada the previous weekend, Wiseco Yamaha rider Tom Kipp, suffered electrical problems and was credited with a 21-25 score

Fast by Ferracci Ducati 851 rider Dale Quarterley was overweight (he and the bike) as well as underpowered in comparison to the 888cc factory Ducati's and the jockey-like European riders, and would retire from the first leg with overheating problems and

sit out the second.

Also missing from the Brainerd race was the Canadian duo of Miguel DuHamel and Steve Crevier. DuHamel was racing in Japan and Crevier was unable to repair the Yamaha engine that he damaged in leg two of the Mosport race after finishing fifth in leg one though he did race the Castrol 250cc grand prix on his Team Ciampini Aprilia.

The venue for the race was the 3 mile, ten-turn Brainerd circuit, which, unlike the previous weekend's track-Mosport, was liked by the riders. Brainerd is a drag strip with a long front straight leading to a flat-out-in-top-gear right-hander, which then winds through some level twisties on the way back to the start. The layout, though rather featureless, is reportedly quite enjoyable to ride, and the Europeans were equally impressed by the generous (for the most part) run-off areas.

Though rain threatened late on race day, the weather remained perfect, warm and dry, though the increasing clouds and afternoon humidity wreaked havoc on some tuners' jetting selections. The situation was especially difficult for Honda riders such as Mertens and Roger Burnett, who claimed that their carburettors are especially sensitive to changes in weather.

### *ENTRY/QUALIFYING*

The biggest surprise in qualifying was not that Merkel set the quickest time (1.43.576 at 104.270 mph), but the circumstances he set that time under. Merkel's equipment was held up in US customs due to paperwork snafus, and in fact he practised on Friday with borrowed helmet (with 'John' painted across the back) and leathers, prompting a

*Belgian Stephane Mertens began to show his championship aspirations with 3-2 placings at Brainerd on his d'Hollander prepared HRC Honda.*

number of spectators to ask who the test rider was on the number one motorcycle. Merkel's two bikes finally arrived, but his crew was without tools and spares until everything else arrived on Saturday.

Just slower than Merkel came Roche (1.43.903), the only other riders to break the 1.44 barrier. Pirovano led the cluster grouped on the next second (1.44.507), followed by Loctite Yamaha-riding Brit Terry Rymer (1.44.656), Bimota's Italian Giancarlo Falappa (1.44.862) and Mertens (1.44.945) the last qualifier under 1.45.

Baldwin did an about-face at Brainerd, switching from the Michelin radial front tyre he used in Canada to a bias-ply, and qualified tenth (1.45.649). He reported that the bias-ply was much more predictable: it's either go slow with the radial or run with the pack on the bias-ply,' he said. The choice was obvious.

Canadian National Superbike Champion Rueben McMurter, second overall at the Canadian round, suffered a main bearing failure in his Honda RC30 late on Friday, and would miss Saturday's first timed session while his crew replaced his motor with one borrowed from a fellow Canadian's production racer. Despite the painfully stock condition of the unit, McMurter posted a fast lap at 1.46.235 to qualify 12th. They would re-build the motor to Superbike specifications over-night.

Brit Burnett crashed in Saturday's practice, separating his shoulder, while Denmark's Rene Rasmussen parted company with his Pepsi-Suzuki in a big way on Friday, and afterwards sported a neck brace in addition to the bandages on his previously broken collarbone both were relegated to spectating on Sunday.

### *RACE 1*

Roche's Italian team mate, Baldassarre Monti, led the first lap, but at the end of lap two slowed suddenly on the front straight with electrical problems, narrowly avoiding being hit by Falappa in the process.

Rymer snatched the lead and led until lap seven when his Dunlop tyres cried "enough". Pirovano then led, chased by Falappa, but the two ex-motorcrossers swapped positions on lap 11. Falappa led just a single lap before falling back to fourth, then falling off in turn six on lap 15 when the Yamaha-powered Bimota spit him off. One of the fuel injection nozzles reportedly stuck in the open position - essentially the 21st century equivalent of a stuck throttle.

Pirovano then led, but Mertens took the lead on lap 13 and ran out front until Roche caught and passed him on lap 16 via a daring, inside pass in turn one. Mertens was having a problem with fuel delivery, and later speculated that it was caused by a faulty petcock.

Once out in front, Roche circulated unchallenged, taking the win by a margin of 0.692 of a second. Pirovano held on to the runner-up placing, with Mertens and Merkel not far in arrears. Rymer toured home in fifth, some distance behind the lead pack, with Monti similarly settling for sixth. Baldwin finished seventh, the victim of a poor tyre choice (not his) and an off track excursion. Anders Andersson (Yamaha) took eighth, Canada's Tommy Douglas (Yamaha) ninth and Arnaiz tenth. McMurter ran as high as fifth early on, but began running out of fuel near the end and finished 11th, climbing to tenth following Arnaiz' disqualification.

Paul Iddon's Yamaha OWO1's motor blew up, in the process spraying Gary Goodfellow's visor with oil, and forcing the Englishman to miss the second leg. Likewise, Finland's Jari Suhonen, whose Yamaha also gave up with mechanical problems. Switzerland's Wolfgang von Muralt fell on lap six but make the leg two start.

### *RACE 2*

Falappa led lap one but pitted on lap two with further engine management system problems - the bike running on just two cylinders when he brought it in. America's best hope, Richard Arnaiz, also dropped out on lap two with an overheating engine.

Roche led lap two, with Rymer, Mertens, Pirovano, Monti and Merkel all running nose-to-tail. McMurter followed Baldwin in eighth, but coming out of turn three McMurter crashed out of the race unhurt.

"Baldwin was on the move and I thought, 'I'm following him'," explained McMurter. "I lost the front end on the pavement change."

Merkel made quick work of Monti and Pirovano to move to fourth, a position he gave up in favour of the lead in lap six, yet would end up holding at the chequered flag.

That lap saw Roche's self-coined "big mistake", when he downshifted one too many times and dropped to sixth behind Merkel, Mertens, Rymer, Baldwin and Pirovano. Not to be outclassed, Roche re-took the lead on lap 16, holding Mertens and Merkel at bay till the end.

Baldwin was running around in fourth when what could have been his best World Superbike finish was stamped out by a faulty clutch. "The clutch was slipping on the sighting and warm-up laps," he said. "Then, something locked up."

Pirovano finished a lonely fourth with an equally singular Rymer fifth. Monti placed sixth, Andersson seventh, Biliotti eighth, Douglas ninth and Goodfellow tenth. Italian Davide Tardozzi, who'd had clutch problems on his Bimota in leg one, and fuel injections problems all weekend, finished 11th after battling with Goodfellow most of the way.

Despite his disappointment in not winning his home country's race, Merkel made no excuses: "These guys are making it really hard for me in my own back yard," he said from the winner's rostrum, this time under an American flag (in Canada, an Italian flag was displayed in honour of his Italian team). "I really wanted to win here in front of the American fans."

The Shopping Channel
DON NESS RACECRAFT
Quality Race Cars & Components
ERNATIONAL R
HONDA WORLD CLASSIC
WORLD SUPERBIKE CHAMPIONSHIP
MICHELIN
PEPSI
YAMAHA
BYRD
MICHELIN
YAMAHA
2
MICHELIN
27
DUCATI
DESMO
DUCATI 851
AGIP

# Round 4

# Brainerd, United States

## 11 June

# Race Data

## Qualifiers

| POS | RIDER/NATIONALITY | ENTRANT/BIKE | TIME |
|---|---|---|---|
| 1. | Fred Merkel (USA) | RCM/Rumi Honda RC30 | 1m 43.57 |
| 2. | Raymond Roche (F) | Squadra Corse Ducati | 1m 43.90 |
| 3. | Fabrizio Pirovano (I) | Belgarda Yamaha FZR750R | 1m 44.50 |
| 4. | Terry Rymer (GB) | Loctite Yamaha FZR750R | 1m 44.65 |
| 5. | Giancarlo Falappa (I) | Bimota Spa YB4EI | 1m 44.86 |
| 6. | Stephane Mertens (B) | Team Mediares Honda RC30 | 1m 44.94 |
| 7. | Roger Burnett (GB) | HRC Honda RC30 | 1m 45.14 |
| 8. | Baldassarre Monti (I) | Squadra Corse Ducati 851S | 1m 45.26 |
| 9. | Jari Suhonen (SF) | Arwidson Yamaha FZR750R | 1m 45.63 |
| 10. | Mike Baldwin (USA) | Bimota Spa YB4EI | 1m 45.64 |
| 11. | Gary Goodfellow (CDN) | Don Knit/Yoshimura Suzuki | 1m 45.92 |
| 12. | Rueben McMurter (CDN) | Action Accessories Honda | 1m 46.23 |
| 13. | Anders Andersson (S) | Yamaha Sweden FZR750R | 1m 46.35 |
| 14. | Richard Arnaiz (USA) | Yamaha FZR750R | 1m 46.48 |
| 15. | Tommy Douglas (CDN) | Team Arom Honda RC30 | 1m 46.71 |
| 16. | Fabio Biliotti (I) | Bimota Spa YB4EI | 1m 46.74 |
| 17. | Davide Tardozzi (I) | Speed Shadows/Bimota | 1m 46.81 |
| 18. | Wolfgang von Muralt (CH) | Honda Schweiz RC30 | 1m 47.61 |
| 19. | Paul Iddon (GB) | Pee Jay Yamaha FZR750R | 1m 47.63 |
| 20. | Rene Delaby (L) | Honda RC30 | 1m 47.73 |
| 21. | Dale Quarterley (USA) | Ducati 851S | 1m 47.83 |
| 22. | Marino Fabbri (I) | Bimota | 1m 47.90 |
| 23. | Doug Chandler (USA) | Kawasaki ZXR750R | 1m 48.77 |
| 24. | Scott Russell (USA) | Suzuki GSXR750R | 1m 49.19 |
| 25. | Tom Kipp (USA) | Yamaha FZR750R | 1m 49.67 |
| 26. | David Sadowski (USA) | Suzuki GSXR750R | 1m 49.90 |
| 27. | Carry Andrew (USA) | Suzuki GSXR750R | 1m 50.23 |
| 28. | Todd Strang (USA) | Yamaha FZR750R | 1m 50.49 |
| 29. | Mike Harth (USA) | Suzuki GSXR750R | 1m 50.62 |
| 30. | Ottis Lance (USA) | Suzuki GSXR750R | 1m 50.95 |
| 31. | Johan v Vaerenbergh (B) | Kawasaki ZXR750R | 1m 51.04 |
| 32. | Eric Moe (USA) | Suzuki GSXR750R | 1m 51.29 |
| 33. | John Jacob (USA) | Yamaha FZR750R | 1m 51.51 |
| 34. | John Eidenberger (USA) | Suzuki GSXR750R | 1m 51.61 |
| 35. | Ricky Orlando (USA) | Suzuki GSXR750R | 1m 52.29 |
| 36. | Jeff Harder (USA) | Suzuki GSXR750R | 1m 56.45 |

**Race Data:** 2 x 20 laps of 3.0 mile circuit
**Total Distance:** 120 miles
**Weather:** Sunny/Cloudy
**Fastest Lap:**
**Existing Lap Record:** Establishes new class record
***No further information supplied by race organisers***

# Results Race 1

| POS | RIDER | LAPS | TIME | MPH | KPH |
|---|---|---|---|---|---|
| 1. | Roche | 20 | 35m 08.37 | 102.45 | |
| 2. | Pirovano | 20 | | | |
| 3. | Mertens | 20 | | | |
| 4. | Merkel | 20 | | | |
| 5. | Rymer | 20 | | | |
| 6. | Monti | 20 | | | |
| 7. | Baldwin | 20 | | | |
| 8. | Andersson | 20 | | | |
| 9. | Douglas | 20 | | | |
| 10. | McMurter | 20 | | | |
| 11. | Biliotti | 20 | | | |
| 12. | Goodfellow | 20 | | | |
| 13. | Fabbri | 20 | | | |
| 14. | Delaby | 20 | | | |
| 15. | Russell | 19 | | | |
| 16. | Andrew | 19 | | | |
| 17. | v Vaerenberg | 19 | | | |
| 18. | Lance | 19 | | | |
| 19. | Jacob | 19 | | | |
| 20. | Moe | 19 | | | |
| 21. | Kipp | 18 | | | |
| 22. | Iddon | 17 | | | |
| 23. | Tardozzi | 17 | | | |

**NOT CLASSIFIED**
Falappa 15 laps,Orlando 10, Strang 8, Quarterley 7, Suhonen 6, von Muralt 6, Harth 5, Harder 4, Sadowski 1, Burnett 0, Chandler 0, Eidenberger 0

# Results Race 2

| POS | RIDER | LAPS | TIME | MPH | KPH |
|---|---|---|---|---|---|
| 1. | Roche | 20 | 35m 06.25 | 102.55 | |
| 2. | Mertens | 20 | | | |
| 3. | Merkel | 20 | | | |
| 4. | Pirovano | 20 | | | |
| 5. | Rymer | 20 | | | |
| 6. | Monti | 20 | | | |
| 7. | Andersson | 20 | | | |
| 8. | Biliotti | 20 | | | |
| 9. | Douglas | 20 | | | |
| 10. | Goodfellow | 20 | | | |
| 11. | Tardozzi | 20 | | | |
| 12. | Chandler | 20 | | | |
| 13. | Delaby | 20 | | | |
| 14. | Fabbri | 20 | | | |
| 15. | von Muralt | 20 | | | |
| 16. | Sadowski | 20 | | | |
| 17. | Russell | 19 | | | |
| 18. | Andrew | 19 | | | |
| 19. | v Vaerenberg | 19 | | | |
| 20. | Lance | 19 | | | |
| 21. | Jacob | 19 | | | |
| 22. | Harder | 19 | | | |

**NOT CLASSIFIED**
Baldwin 13 laps, Orlando 12, Kipp 6, Harth 5, McMurter 2, Arnaiz 2, Falappa 2,Burnett 0, Suhonen 0, Iddon 0, Quarterley 0, Strang 0, Moe 0, Eidenberger 0

***No further information supplied by race organisers***

# Championship Positions

**1.** Merkel 126, **2.** Pirovano 105, **3.** Rymer 78, **4.** Mertens 77, **5.** Roche 74, **6.** Andersson 69, **7.** Falappa 60, **8.** Monti 40, **9.** McMurter 38, **10.** Burnett 32, **11.** McElnea 28, **12.** Baldwin 27, **13.** Suhonen 23, **14.** Gschwender 22, **15.** Douglas 20, **16.** Rasmussen 18, **17.** Morrisson, Rubatto 17, **19.** Igoa, Biliotti 16, **21.** Goodfellow 15, **22.** Hofmann 14, **23.** Hislop 13, **24.** Fogarty 12, **25.** Crine, Phillips, Crevier 11, **28.** Iddon, Tardozzi 8, **30.** Mercier 7, **31.** Campbell, McGladdery, Delaby, Fabbri 5, **35.** Chandler 4, **36.** von Muralt 3, **37.** Moineau 2, **38.** Bouheben 1

# Round 5

# Osterreichring, Austria

## 2 July

***(Above)*** *Aussie Malcolm Campbell gets a flier. Others prominent include Pierre Bolle (16), Patrick Igoa (29), Stephane Mertens (4), Fred Merkel (1), Alex Vieira (6) and Baldassare Monti (100).* ***(Top Right)*** *Monti leads Mertens, Broccoli and Pirovano into the Hella Licht.*

Hella-Licht S

FALAPPA fans
FALAPPA fans
FALAPPA fans

World Superbike continued it's unpredictable trend in Austria, the fifth round of the championship. We had two first time winners in Alex Vieira and Stephane Mertens, a superb performance by Superbike 'rookie' Massimo Broccoli on the Ducati and a surprising fifth place in the second leg by Kawasaki France pilote Patrick Igoa.

Defending champion, Fred Merkel retained the point lead after a gutsy second race third but dropped eight points to his nearest rival Fabrizio Pirovano in the overall standings. The little Italian scored a fifth and second in the respective races.

British rider Terry Rymer, third in the series going into the Austrian round, dropped to fifth after a lacklustre performance on his Yamaha due to a shoulder injury.

The weather made life difficult for the competitors. All weekend, a threatening, swirling mist over the mountains had occasionally turned to rain during qualifying and sure enough, the heavens opened late Sunday to make the second race a complete lottery.

But both races were the greatest advert possible for World Superbike. In the first there were up to a dozen riders running nose to tail for the lead and in the second three riders duked it out for the winning place in the worst conditions thinkable, and another pack of eight fought out fourth place.

The biggest disappointment? Only 10,00 people had the pleasure of seeing World Championship motorcycle racing at it's best.

***ENTRY/QUALIFYING***

Emphasising the strength and competitive nature of World Superbike, there were 104 entries for this fifth round of the championship. Seventy seven took part in timed qualifying and fifty riders, separated by just 6.97 seconds, took the green light at the start of race one.

Interesting to note that the same weekend just 28 riders practiced for the World Championship 500cc race at Spa Francorchamps in Belgium.

Dane Rene Rassmussen, the Pepsi Suzuki rider, was absent from Osterreichring, still recovering from the head injuries he suffered in his Brainerd practice crash.

Fabio Billioti was also a no-show. He had quit the Bimota camp on the Tuesday prior to race weekend and there was no time left to fly American ace Mike Baldwin in to take his place. It came as little surprise when the bespectacled Italian had opted out of Superbike. He never adapted to the big fourstrokes and failed to produce a point scoring finish for the team.

The Osterreichring, with only one corner the 65mph second gear 'Hella Licht S' chicane, under 80mph, was expected to favour the speedy Ducatis. And the big vee twins were out in force.

Team leader Raymond Roche grabbed pole with a 1m 55.610 but this came late in the final qualifying session after rain had interrupted the first one.

It was a brave effort by the Frenchman. He had taken a high speed tumble in Friday's untimed training when he came across a couple of national class CBR500s. He spent much of the remainder of the day holding an ice pack on a badly swollen elbow.

New boy Baldasarre Monti was also fast with a 56.100, fourth fastest overall but he would start sixth and on the second row of the grid. The grid was determined by picking the fastest rider alternately from each session rather than fastest overall.

Monti was joined on row two by yet another of the famous fire engine red machines with another Italian, Massimo Broccoli in the saddle, fifth fastest with a 56.183. An experienced hand on all types of GP machinery, Broccoli was relatively new to the thundering vee twins but came into the race full of confidence after scoring an Italian National championship victory at Imola the previous weekend for Marco Lucchinelli's Squadra Corse Ducati team.

However, only two other riders besides Roche broke the 1.55 barrier and both were Honda mounted. Alex Vieira made a

*Giancarlo Falappa gingerly leads the pack through th*
*wasn't to last. He threw it away going into the same c*

*lla Licht chicane early in the wet second race. It r on the third lap.*

welcome return to the Superbike ranks and posted a 55.610 on the Honda France entry while Belgian Stephane Mertens did a 55.970 with his Jean d'Hollander fettled HRC machine which runs in the colours of Team Mediares/Total/Bel Ray, Mediares being a Belgian based promotions company.

After the two other Ducatis came Fred Merkel on another of the special HRC bikes, his in the purple and black colours of Gypsy, the Italian based underwear and tee-shirt manufacturer. Fred rattled off a 56.290 in his second session but at one stage looked in danger of not qualifying at all.

Timed qualifying was split into two groups, odd and even numbers. It rained early in the session and several riders were not on the pace quick enough to record a fast dry lap. In fact only one rider in the entire session broke the two minute barrier, that rider being Patrick Igoa on the Kawasaki.

The riders went along to ask for a replacement session to run after the even numbers, a request which was granted. The riders were all ready to go out on the dry track when Bimota boss, Giuseppe Morri protested, arguing that his team was not prepared for the hitherto unscheduled session.

So there was no extra practice and Merkel having posted only a 2m 11.400 best was facing a non qualifying slot if it rained in the second session. Only three riders had gone slower among the odd numbers!

Luckily it was dry long enough for Merkel in the second session and he quickly scored a 56.270 which put him second to Roche in their group. After his quick lap Merkel stormed off down the pit lane to confront the Bimota team. He had to be dragged away from their pit garage by his own team members, making loud protestations in his best Italian to the Bimota team boss as he went.

So what did the American star say to the rival Italian team? "Well, I can't tell you exactly but I wanted to make a point that's all. Last year every time they (Bimota) were beaten they would protest everyone in sight and there's no need for it. So I went down there to say hey, your dirty trick backfired on you this time."

Light drizzle hit the second odd numbers session too but not before most riders had managed some respectable times. The average number of laps for the top fifteen in the group was 20.80 compared to 28.46 in the even numbers group who had a pair of dry sessions.

Merkel only completed nineteen laps total in timed qualifying and several of those were touring laps to check out the course. But he did entertain the pit lane crowd with some superb wheelies the entire length of the start line straight.

Just 0.10 of a second slower than Merkel was Malcolm Campbell, the Aussie star having come half way around the world to ride an Austrian prepared bike entered by former racer Walter Adler. The immaculate aqua coloured RC30 with an '89 kit motor may have not been the fastest bike on the track but 'Wally' was certainly flying and lapping about two seconds a up on his '88 times.

His team had a scare in the half hour warm up when 'Wally' had a clutch break up on him, but as the Aussie said it was better to happen then than when he was on the line for the race.

Another 0.10 back came Anders Andersson on the first of the Yamaha OWO1s then came Davide Tardozzi who had former Bimota team manager Federico Martini helping out in the Speed Shadows team garage. Patrick Igoa (Kawasaki France ZXR), and Ernst Gschwender were also within the 1m 56s bracket.

The West German had recorded his fast time on the Suzuki Deutschland GSXR750R by getting a tow from Andersson late in the session but one lap later lost the rear end going into the chicane and crashed heavily. It was later discovered that the oil cooler had burst and coated the rear slick with lubricant.

The bike was not seriously damaged but the lanky Gschwender took a blow to his left thigh. "It was like someone had hit my leg with a hammer,'' he winced. Ignoring the obvious pain, he was on the grid for Sunday's race.

Igoa confirmed that the French Kawasaki team, under the direction of Georges Godier, had spent a lot of hours getting the green meanies competitive. ''We have no special parts from Japan but have done a lot of dyno work. We have found an extra six bhp since Donington Park.''

The bike was now bang on the FIM weight limit and the only real handicap for the riders was the rather lengthy wheelbase which made the bike a bitch to change direction with.

''It is okay on a circuit like Osterreichring,'' revealed the former GP500 runner, ''because the corners are long and fast but at circuits like the Hungaroring where there are lots of quick changes of direction, it is impossible. It is something we are working on for the future.''

Pierre Bolle, still standing in for the injured Christophe Bouheben on the second of the factory Kawasakis, was only 0.5 of a second slower than the Suzuki rider and a fraction quicker than the singleton factory Bimota of Giancarlo Falappa. The exhuberent Italian had brought a coach load of 'Falappa Fans' fans with him but so far was finding it difficult to get on the pace in this his first visit to the 'Ring'.

He was another of the 'crocks' in the line up, carrying a very swollen knee after whacking it on a kerb at Imola and was having some difficulty getting his stocky frame into a racing crouch on the compact machine.

Several top names were languishing further back than many expected. Brit, Terry Rymer was carrying a shoulder injury from two weeks previous but it didn't stop him racking up nearly 200 miles of practice in free training on Friday with the Loctite Yamaha.

Main reason for the high mileage was to get dialled in with some new Dunlop rubber. For the first time all year he had Grand Prix spec tyres, as used by the Roberts/Lucky Strike team to fit his specially made that-week Dymag 17 inch diameter, six inch wide wheels. The aim was to find a tyre capable of lasting a race distance, something the 18" Superbike tyres had not previously delivered.

And his team had some burning of the midnight oil prior to race day. The OWO1 blew a head gasket in the last few laps of timed training necessitating a complete engine removal and rebuild of the top end.

Second in points Fabrizio Pirovano was suffering from a painful neck complaint caused, he suspected, from a draught, and was taking medicine to overcome the discomfort. His lap time on the Belgarda/ Byrd Yamaha was way off what many expected of him, a 57.570.

Roger Burnett was next to Pirovano on the grid but his time was only in the 58's. The HRC rider prefers to rack up several laps in qualifying but the rain interrupted session prevented that. His crew actually clocked only a dozen flying laps out of his twenty total, his best a 58.100. And like Rymer, he too was troubled a shoulder injury, legacy of his Brainerd endo.

The Minolta Suzukis of Suzuki France were again way off the pace. There had been no serious changes to the machinery since the team had fronted at Donington earlier in the year. A great shame considering both Herve Moineau and Thierry Crine are regarded as top flight competitors and not only at Endurance championship level.

Although the team may not have a serious commitment to Superbike, Suzuki would seem to have high hopes for Crine in the future and he was due to test a spare factory RGV500 in readiness for the French GP on the Tuesday/Wednesday after the Osterreichring meeting. Perhaps that might be an indicator to the pony tailed character's lack of fire during the Austrian meeting - or perhaps both him and Moineau were fed up with campaigning an uncompetitive machine.

For sure Crine was frustrated with the Austrian effort. He said, ''you know, at the start of the year I was able to compete with Vieira and Mattioli. Now my bike is three to four seconds a lap slower. Gschwender has worked on his Suzuki and Moineau followed him for some laps but could not keep up on the straights.''

***RACE 1***

It was hot and humid for the opening race and the mist was still hanging in the pine trees on the higher slopes indicating changeable weather conditions in store.

Monti led the charge into the Hella Licht from Mertens but the Belgian soon took control and led the field through at the end of the opening lap with Vieira in tow.

The leading pair had established a break on the bunch which was led by Monti from Tardozzi, Merkel, Campbell, Broccoli, and a fast starting Suhonen. Then there was a brief gap to Roche, Rymer, Falappa, Pirovano, Gschwender and Bolle.

Andersson had made a lousy start and was 18th but charging through on the Yamaha. Igoa had equally suffered a bad start. His clutch went bang on the line and he was forced to fight a rear guard action way down the field.

By lap three the third place chasing pack had latched onto the two leaders who had been slowing each other with their own private scrap. A two man dice had suddenly become a twelve man dice for the lead and the small crowd were going wild to breathe some sort of atmosphere into the event.

On lap five Tardozzi made his bid for the lead, a move which sent the Falappa fans into a frenzy. Their own hero was among the lead bunch but for the first time in the race they saw an Italian in front and that was a moment they wanted to savour.

It was a brief lead for the normally excitable Italian. Two laps later Tardozzi swung the yellow YB4 into pit row to retire with a suspected main bearing failure. After all his previous engine problems this year, he had opted to run an '88 motor but the gremlins had struck once again.

Still the fans were cheering though as

Monti, clocking low 1m 55s made a bid for the lead on the inside under braking for the chicane, swapping places with Mertens at the front of a twelve rider freight train. Broccoli was also prominent as were Vieira, Pirovano, Roche (another to have made a slow getaway), Merkel, Falappa, Rymer, and Campbell.

Falappa had come through from a lowly start and had been clocking 1m 55 laps but then started to have problems with his Bimota. Intermittently, it refused to rev out but for no apparent reason which meant his position on the leaderboard was erratic to say the least.

Andersson was still plugging away to get on terms and had Suhonen right on his tail. Then came a lone Igoa followed by another slipstreaming bunch which included his Kawasaki teammate Bolle, Michel Simul on a Van Zuylen MotorsportsRC30, Gschwender, Delaby and Mattioli.

At mid distance it was Broccoli's turn to lead and one couldn't help thinking there were Ducati team orders to slow the pack until team leader Roche got on terms with the bunch.

But Monti was looking pretty desperate to stay among the front runners as he repeatedly got out of shape trying to forge past Vieira.

Slowly but surely Roche moved through the pack and by lap eleven was sitting on the tail of Mertens, Broccoli and Vieira. Pirovano and Monti were also in contention but a slight gap was now appearing with Falappa now running on his own from Merkel who was in tyre trouble with the Pirelli, Rymer in trouble with his shoulder, Campbell, Suhonen and Andersson.

The latter in his early charging laps, with several 1m 55s thrown in, had cooked the rear tyre and was now 'sitting in' on the action to conserve the rubber and give himself a breather after his earlier exertions.

Moineau retired the second Suzuki with

*Malcolm Campbell was in great form on the Austrian supplied Walter Adler Honda.*

*Vieira leads Mertens during their first race clash.*

'motor problems' to join Crine in pit lane.

But the leaders still hadn't finished and all round the circuit they were jockeying for position. Lap thirteen nearly brought the house down. Well, it would have done if the crowd had been bigger. But those who were there cheered as Broccoli and Monti formed a Ducati 1-2 at the front.

Broccoli held station until lap sixteen when Roche went inside at the chicane to take over. It was as many expected. The wily GP veteran had been bidding his time and would now control the race for the final two laps.

Or would he? Even coming out of the long left called Schikane on the final lap the Ducati led but as they took the uphill rise toward the final Rindt Kurve, Vieira forged by on the Honda and kept ahead on the final run in to the line for a dramatic victory.

Roche kept second from Mertens who had suspension problems which overheated the rear tyre, Broccoli, Pirovano, and Monti - the group split by 1.331 seconds - five bike lengths covering six riders!

A tad further back came Campbell then another 1.5 seconds saw Andersson. Rymer was three seconds further adrift with Suhonen, and Merkel right with him. It was an unhappy Merkel who brought the Rumi/ Gypsy Honda in pit lane afterwards. He parked up, got off then ran his gloved hand over the rear tyre, shaking his head as he did so, then walked off. There was no need for words - his actions said it all.

Igoa was a lone thirteenth, Falappa - lucky to see the flag with his Bimota firing intermittently towards the end - fourteenth and Bolle rounded out the point scorers.

***RACE 2***

The start of the second race was meant to have been at 3.15 but the demonstration of a jet powered drag car and a couple of drag bikes meant the start was delayed. The dragsters were meant to have been at 11.15 but they weren't ready and they were still fiddling around in pit row come the time we should have seen the Superbikes on the grid.

The organisers gave the dragsters priority because they were using the demonstration as a promotion for their big drag meeting a few weeks ahead at the same venue. It was a pity no one could have intervened to get the real racing underway on time.

It was to prove a costly delay because as the bikes started their sighting lap at 3.25, it began to rain. The riders set off on their warm up lap at 15.40 but as soon as they came round the officials had to red flag the race before it started to enable the teams time to fit wet tyres.

Nearly an hour later, with the weather conditions even worse, the riders set off on a warm up lap and the race was reduced to fifteen laps.

Fred Merkel led the field around for the first lap, with everyone tiptoeing around the flooded and very treacherous track.

Lap two saw Roche take up the running as Merkel slipped down to sixth. Also in the frame were Pirovano, Mertens and Eric Delcamp on his privately entered Dubois Motos of Nimes OWO1. Then came Merkel, Tardozzi, Bolle, Vieira, Andersson, Broccoli, Fabbri, Monti who had incurred a one minute penalty for jumping the start, and Falappa.

Pirovano slithered into the lead at the Hella Licht while Falappa threw it away at the same spot. His Bimota aquaplaned when he applied too much force on the Brembos and he slid off at the crest of the rise on the straight ending up in the tyre barrier marking the run off.

Unhurt, he jumped up pulled off his Nava helmet and took a grand applause from his own Falappa Fans who were relieved to see their hero unhurt.

The conditions were taking their toll. Burnett retired in pit lane, his shoulder hurting from every little wrench on the bars induced by slides on the wet track. Two laps later Rymer joined him on the retirement list with the same complaint.

Back on the track, Pirovano still led but had spilt the pack and now only had Mertens in his wheel tracks with Merkel chasing hard - given the severity of the conditions. Pirovano was lapping at around 2m 21.88, when the conditions were at their worst.

The group behind consisted of Roche, Gschwender, Broccoli, Vieira, Monti, Delcamp, Campbell, Igoa, and Andersson. Then another gap to Tardozzi and Bimota rider Marino Fabbri who was later to crash.

Roche was trying to bridge the gap to Merkel and on lap six it was undoing. He went in too hot into the chicane, lost the front end and had to pick everything up and head for the slip road.

By the time he wound through the tyre barriers and rejoined the circuit he was at the back of the pack he once led. But after the incident he clearly decided to opt for safety and slipped further downfield out of the serious action.

Pre-race speculation suggested the Ducatis would have problems in a wet race with their high compression motors locking the rear wheel in downshifts but no one told Broccoli or Monti who kept up a fierce pace at the head of the chasing fourth place group.

The following lap Merkel did exactly the same as Roche and went up the slip road. He recovered without losing a place and set off after the two leaders again. Two laps later he got out of shape in the middle of the chicane and required some frantic footwork to stay aboard.

Merkel later said, "I was trying really hard and the chicane was one of my good points on the race track. But then it started raining real hard and got real slippery - as everyone saw - so I was just kinda hanging on."

Mertens took over the lead on the ninth lap and managed to hold on to the finish from Pirovano while Merkel had to settle for a hard fought third.

Broccoli led the bunch sprint out of the Rindt Kurve to claim fourth, Igoa was a very surprised fifth - the best result for Kawasaki in the series so far. "I do not go fast in the wet," he beamed, " so this result is a surprise for me also."

Monti finished in sixth place but his one minute penalty dropped him to 14th. That moved Campbell up to sixth from Suhonen, Vieira and Andersson.

Campbell was delighted with his two point scoring rides on the day, especially the second race. "That was a bit of a goer eh?" he grinned. "I though I had got rid of Vieira and Igoa but they came back at me with two to go. But it was an incredible race for the conditions. I really enjoyed it."

After Andersson, the field was spread out with Roche a distant tenth, Gschwender, Tardozzi, Ricci, the luckless Monti and Delaby rounding out the point finishers.

After his first Superbike win of the season Mertens said, "it was a problem just to stay on two wheels and very hard to stay with Pirovano and Merkel. But Merkel went straight on at the chicane and then I stayed with Pirovano.

"I thought I could win but it was not easy. At the chicane Pirovano was faster on acceleration but I was very fast at the Rindt Kurve. So I knew I needed to pass him before the chicane and then I would be okay. It was easy to keep in front of him in the other parts of the track."

# Qualifiers

# Round 5

# Osterreichring, Austria

## 2 July

# Race Data

| POS | RIDER/NATIONALITY | ENTRANT/BIKE | TIME |
|---|---|---|---|
| 1. | Raymond Roche (F) | Squadra Corse Ducati 851S | 1m 55.38 |
| 2. | Alex Vieira (P) | Honda Motul 750 | 1m 55.61 |
| 3. | Stephane Mertens (B) | Total Bel Ray Honda 750 | 1m 55.97 |
| 4. | Baldassarre Monti (I) | Squadra Corse Ducati 851S | 1m 56.10 |
| 5. | Massimo Broccoli (I) | Squadra Corse Ducati 851S | 1m 56.18 |
| 6. | Fred Merkel (USA) | RCM/Rumi Honda RC30 | 1m 56.27 |
| 7. | Malcolm Campbell (AUS) | Citizen Honda RC30 | 1m 56.28 |
| 8. | Anders Andersson (S) | Yamaha FZR750R | 1m 56.29 |
| 9. | Davide Tardozzi (I) | Bimota YB4EIR | 1m 56.62 |
| 10. | Patrick Igoa (F) | Team Kawasaki France | 1m 56.63 |
| 11. | Ernst Gschwender (D) | Suzuki GSXR750R | 1m 56.71 |
| 12. | Pierre Bolle (F) | Team Kawasaki France | 1m 56.76 |
| 13. | Giancarlo Falappa (I) | Bimota Spa YB4EI | 1m 57.13 |
| 14. | Terry Rymer (GB) | Team Loctite Yamaha OWO1 | 1m 57.14 |
| 15. | Rene Delaby (B) | Honda RC30 | 1m 57.16 |
| 16. | Fabrizio Pirovano (I) | Yamaha FZR750R | 1m 57.17 |
| 17. | Marino Fabbri (I) | Bimota YB4EI | 1m 57.19 |
| 18. | J. Michel Mattioli (F) | Honda Motul 750 | 1m 57.27 |
| 19. | Jari Suhonen (SF) | Arwidson Yamaha FZR750R | 1m 57.63 |
| 20. | Paul Iddon (GB) | Yamaha FZR750R | 1m 57.68 |
| 21. | Peter Rubatto (D) | Hein Gericke Bimota YB4EI | 1m 57.87 |
| 22. | Herve Moineau (F) | Minolta Suzuki GSXR750R | 1m 57.95 |
| 23. | Michel Simul (B) | Van Zuylen Honda RC30 | 1m 57.95 |
| 24. | Thierry Crine (F) | Minolta Suzuki GSXR750R | 1m 58.05 |
| 25. | Roger Burnett (GB) | HRC Honda RC30 | 1m 58.10 |
| 26. | Eric Delcamp (F) | Yamaha FZR750R | 1m 58.44 |
| 27. | Michael Galinski (D) | Mitsui Yamaha FZR750R | 1m 58.57 |
| 28. | Mauro Ricci (I) | Ducati 851 | 1m 58.74 |
| 29. | Antonio Garcia (E) | Gabi Comp Honda RC30 | 1m 58.92 |
| 30. | Karl-Heinz Riegl (A) | D. Freythe Bimota YB4EIR | 1m 59.16 |
| 31. | Jeffrey de Vries (NL) | Yamaha FZR750R | 1m 59.32 |
| 32. | Anton Heiler (D) | Bimota YB4EIR | 1m 59.34 |
| 33. | Hans Lindner (A) | Honda RC30 | 1m 59.39 |
| 34. | Wolfgang von Muralt(CH) | Honda Schweiz RC30 | 1m 59.64 |
| 35. | Arnad Harmati (H) | Honda VFR750R | 1m 59.65 |
| 36. | Harald Oberprank (A) | Honda Bimmer RC30 | 1m 59.73 |
| 37. | Anton Gruschka (D) | Honda RC30 | 1m 59.75 |
| 38. | Karl Katoch (A) | Team Kawasaki (A) ZXR750 | 2m 00.35 |
| 39. | Johann Parzer (A) | Yamaha OWO1 | 2m 00.40 |
| 40. | Michael Schon (A) | Honda Mayerhofer RC30 | 2m 00.74 |

| | |
|---|---|
| **Race Data:** | 2 x 18 laps of 3.63 mile circuit |
| **Total Distance:** | 130.68 miles |
| **Weather:** | Dry, cloudy/raining |
| **Fastest Lap:** | Baldassare Monti (race one) 1m 55.195s, 113.53 mph, 182.822 kph |
| **Existing lap record:** | Marco Lucchinelli 1m 56.594s, 112.22 mph, 180.718 kph (July 1988) |

# Results Race 1

| POS | RIDER | LAPS | TIME | MPH | KPH |
|---|---|---|---|---|---|
| 1. | Vieira | 18 | 34m 58.39 | 112.24 | 180.75 |
| 2. | Roche | 18 | 34m 58.57 | 112.23 | 180.73 |
| 3. | Mertens | 18 | 34m 59.01 | 112.21 | 180.69 |
| 4. | Broccoli | 18 | 34m 59.20 | 112.20 | 180.68 |
| 5. | Pirovano | 18 | 34m 59.55 | 112.18 | 180.65 |
| 6. | Monti | 18 | 34m 59.72 | 112.17 | 180.63 |
| 7. | Campbell | 18 | 35m 00.04 | 112.15 | 180.60 |
| 8. | Andersson | 18 | 35m 01.59 | 112.07 | 180.47 |
| 9. | Rymer | 18 | 35m 04.63 | 111.91 | 180.21 |
| 10. | Suhonen | 18 | 35m 04.79 | 111.90 | 180.20 |
| 11. | Merkel | 18 | 35m 05.46 | 111.87 | 180.14 |
| 12. | Igoa | 18 | 35m 08.23 | 111.72 | 179.90 |
| 13. | Falappa | 18 | 35m 15.42 | 111.34 | 179.29 |
| 14. | Bolle | 18 | 35m 20.80 | 111.07 | 178.86 |
| 15. | Gschwender | 18 | 35m 21.59 | 111.01 | 178.76 |
| 16. | Iddon | 18 | 35m 21.76 | 111.01 | 178.76 |
| 17. | Mattioli | 18 | 35m 22.75 | 110.95 | 178.67 |
| 18. | Delaby | 18 | 35m 23.18 | 110.93 | 178.63 |
| 19. | Burnett | 18 | 35m 25.98 | 110.79 | 178.40 |
| 20. | Delcamp | 18 | 35m 27.73 | 110.70 | 178.25 |
| 21. | von Muralt | 18 | 35m 31.69 | 110.49 | 177.92 |
| 22. | Fabbri | 18 | 35m 50.90 | 109.50 | 176.33 |
| 23. | Lindner | 18 | 36m 00.48 | 108.91 | 175.38 |
| 24. | de Vries | 18 | 36m 00.87 | 108.87 | 175.32 |
| 25. | Seidel | 18 | 36m 04.39 | 108.82 | 175.23 |
| 26. | Ricci | 18 | 36m 04.74 | 108.80 | 175.20 |
| 27. | Galinski | 18 | 36m 12.06 | 108.44 | 174.62 |
| 28. | Parzer | 18 | 36m 17.60 | 108.16 | 174.17 |
| 29. | Heiler | 18 | 36m 17.72 | 108.15 | 174.16 |
| 30. | Gruschka | 18 | 36m 18.58 | 108.11 | 174.09 |
| 31. | Ommer | 18 | 36m 18.80 | 108.10 | 174.08 |
| 32. | Bontempi | 18 | 36m 19.48 | 108.07 | 174.02 |
| 33. | Pajic | 18 | 36m 38.05 | 107.15 | 172.55 |
| 34. | Franz | 18 | 36m 38.36 | 107.14 | 172.52 |
| 35. | Mitter | 18 | 36m 38.91 | 107.13 | 172.51 |
| 36. | Schon | 18 | 36m 39.24 | 107.10 | 172.46 |
| 37. | Meier | 18 | 36m 39.56 | 107.08 | 172.43 |
| 38. | Benedetti | 18 | 36m 40.43 | 107.04 | 172.36 |
| 39. | Schlafli | 16 | 35m 05.60 | 99.43 | 160.11 |
| 40. | Hager | 16 | 35m 14.94 | 98.99 | 159.41 |
| 41. | Krummenacher | 15 | 35m 26.12 | 92.31 | 148.66 |

# Results Race 2

| POS | RIDER | LAPS | TIME | MPH | KPH |
|---|---|---|---|---|---|
| 1. | Mertens | 15 | 35m 28.94 | 92.19 | 148.46 |
| 2. | Pirovano | 15 | 35m 29.95 | 92.15 | 148.39 |
| 3. | Merkel | 15 | 35m 35.88 | 91.89 | 147.98 |
| 4. | Broccoli | 15 | 35m 40.50 | 91.70 | 147.66 |
| 5. | Igoa | 15 | 35m 41.20 | 91.67 | 147.61 |
| 6. | Campbell | 15 | 35m 42.74 | 91.60 | 147.50 |
| 7. | Suhonen | 15 | 35m 45.17 | 91.50 | 147.34 |
| 8. | Vieira | 15 | 35m 45.32 | 91.49 | 147.33 |
| 9. | Andersson | 15 | 35m 47.75 | 91.39 | 147.16 |
| 10. | Roche | 15 | 36m 02.95 | 90.74 | 146.13 |
| 11. | Gschwender | 15 | 36m 10.55 | 90.43 | 145.61 |
| 12. | Tardozzi | 15 | 36m 22.08 | 89.95 | 144.85 |
| 13. | Ricci | 15 | 36m 38.33 | 89.28 | 143.77 |
| 14. | Monti | 15 | 36m 42.59 | 89.11 | 143.50 |
| 15. | Delaby | 15 | 37m 06.36 | 88.16 | 141.96 |
| 16. | Iddon | 15 | 37m 08.76 | 88.06 | 141.81 |
| 17. | Bolle | 15 | 37m 12.52 | 87.92 | 141.57 |
| 18. | Galinski | 15 | 37m 26.64 | 87.36 | 146.82 |
| 19. | Bontempi | 15 | 37m 32.13 | 87.15 | 140.34 |
| 20. | Meier | 15 | 37m 33.91 | 87.08 | 140.23 |
| 21 | Gruschka | 15 | 37m 45.55 | 86.63 | 139.51 |
| 22. | von Muralt | 15 | 37m 46.88 | 86.58 | 139.43 |
| 23. | de Vries | 15 | 37m 48.86 | 86.51 | 139.30 |
| 24. | Heiler | 15 | 37m 52.47 | 86.37 | 139.08 |
| 25. | Franz | 14 | 35m 34.87 | 85.81 | 138.18 |
| 26. | Parzer | 14 | 35m 58.33 | 84.88 | 136.68 |
| 27. | Schlafi | 14 | 36m 10.69 | 84.39 | 135.90 |
| 28. | Schon | 14 | 37m 10.39 | 82.13 | 132.26 |
| 29. | Pajic | 13 | 36m 25.84 | 77.82 | 125.32 |

**NOT CLASSIFIED (RACE 1)**

Garcia 13, Riegl 13, Simul 13, Moineau 9, Tardozzi 7, Oberprank 7, Crine 4, Katoch 1, Rubatto 1

**NOT CLASSIFIED (RACE 2)**

Fabbri 11 laps, Lindner 8, Delcamp 8, Katoch 7, Garcia 6, Mattioli 4, Krummenacher 4, Rymer 3, Seidel 2, Falappa 2, Benedetti 1, Burnett 1, Hager 1

# Championship Positions

**1.** Merkel 146, **2.** Pirovano 133, **3.** Mertens 112,**4.** Roche 97, **5.** Rymer 85, **6.** Andersson 84, **7.** Falappa 63, **8.** Monti 52, **9.** McMurter , Suhonen 38, **11.** Burnett 32, **12.** Igoa 31, **13.** McElnea, Vieira, Gschwender 28, **16.** Baldwin 27, **17.** Broccoli 26, **18.** Campbell 24, **19.** Douglas 20, **20.** Rasmussen 18, **21.** Biliotti, Morrison, Rubatto 17, **24.** Goodfellow 16, **25.** Hofmann 14, **26.** Hislop 13, **27.** Fogarty, Tardozzi 12, **29.** Crine, Phillips, Crevier 11, **32.** Iddon 8, **33.** Mercier, Delaby 7, **35.** Fabbri 6, **36.** McGladdery 5, **37.** von Muralt, Ricci 3, **39.** Moineau, Bolle 2, **41.** Bouheben, Russell 1

# Round 6

## Paul Ricard, France

30 July

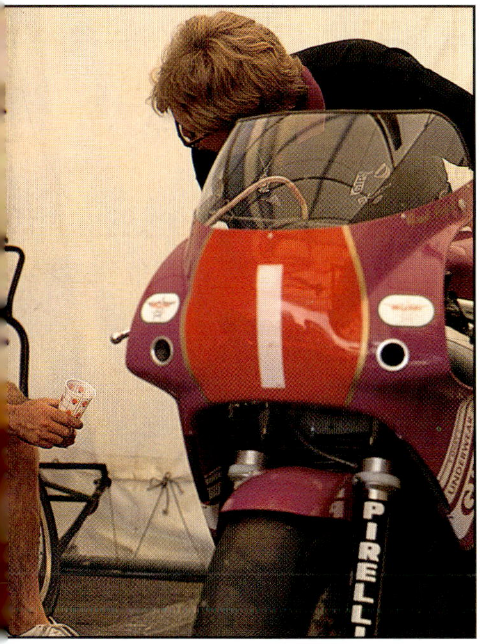

*(**Above**) Chief Pirelli tyre technician Peter Ingley in deep discussion with Fred Merkel's crew. (**Left**) Mike Baldwin.*

Belgian Stephane Mertens piled on the pressure in the sixth round of the World Superbike championship, held at Paul Ricard in Southern France, to close to within twenty points of series leader, and defending champion, Fred Merkel.

With a display of hard, calculated riding, Mertens, riding the Jean D'Hollander prepared HRC supplied Honda, won one leg and finished a fighting third in the other.

For Californian Fred Merkel it was a disappointing day, with his similar to Mertens' HRC Honda, beset with minor mechanical problems and both he and his Oscar Rumi/Gypsy team members knew that he faced a tough battle to keep the on form Mertens at bay.

Meanwhile at the Bimota camp, some form of consistency seemed to have been found with the continued presence of experienced American Mike Baldwin in the team. With the aggressive Italian Giancarlo Falappa riding less erratically, and staying on board, Bimota had become a real force to be reckoned with as the season progressed.

Falappa was riding just as hard at Ricard, it is just that some of the rough edges had been polished off as he gained experience. It was, after all, his first season of international racing!

Falappa and his compatriot, Fabrizio Pirovano both come from a motocross background, though the previous knowledge did not help poor Pirovano in the first race when he lost the front end and went down

*Stephane Mertens, flanked by Mike Baldwin (left) an*
*Roche wanted a win in front of his home crowd but h*

heavily. He didn't finish the second race either and was probably wishing he had stayed home to allow the wrist he broke at Misano a week earlier in an Italian Superbike round, a longer period to heal.

***ENTRY/QUALIFYING***

As usual, qualifying for a place on the grid

*aymond Roche on the rostrum, won one and placed third in the other race at Ricard to pressure points leader Fred Merkel.* ***(Left)*** *Raymond*
*settle for rostrum places.*

was as hectic as ever. The track itself was still slippery from the Formula One car Grand Prix just two weeks previous and together with the intense heat, most riders were having tyre problems.

For the first time in the 1989 series, there was a Greek entrant in the ranks, Satis Skourtas, their national champion who was entered by Rothmans Greece on the ex-Terry Rymer Honda RC30.

Having only previously raced on the Greek airfield tracks, he was shown the way around the course by fellow Rothmans rider, Britain's Roger Burnett and was only a couple of seconds off the pace in free training. However, it was to no avail. an oil line broke on the Honda and Satis crashed in the untimed practice and broke his wrist in two places.

Burnett himself was lucky to escape unhurt from a high speed tumble at the end of the Mistral straight when he lost the front end in the fast right hander. His bike needed a hasty rebuild however.

Local hero Raymond Roche dominated

NOLAN
Castrol
PIRELLI
STO

Marlboro
Marlboro
-POULENC
bimota
76
15
77
1
2
4
90
3

qualifying with the Squadra Corse Ducati 851, despite the effects of a heavy cold and a dislike of the Paul Ricard short circuit. His time of 1m21.83s was just one tenth of a second faster than Merkel. The remainder of the front row was made up by Baldwin (1m 22.16), Monti (1m 21.98) and local bikeshop dealer Maurice Coq who was clearly using his extensive circuit knowledge to his advantage and got down to a 1m 22.45 in the second timed session on his FZR750R Yamaha.

***RACE 1***

With the temperature hitting forty degrees, the riders set off on their warm up lap but as the leading riders arrived back at the grid there was much arm waving, indicating something was seriously wrong. At the end of the Mistral straight several machines had apparently collided and riders reported seeing bikes and bodies catapulted through the air.

It transpired that the Hondas of Belgian Rene Delaby and Swiss Wolfgang von Muralt had collided and in the ensuing crash had also brought down von Muralt's Honda RC30 mounted teammate, Toni Rohrer. Von Muralt broke his right leg and wrist, Delaby suffered concussion and injured his spine while poor Rohrer, unable to miss the debris, broke his right thigh.

The start was delayed by twenty minutes and the re-run was scheduled to be over twenty seven laps. Thirty seven riders blasted off into turn one for the re-run with British Loctite Yamaha rider Terry Rymer leading Monti, Pirovano, Mertens and Merkel.

Another Brit, Roger Burnett didn't even get as far as the first turn. He collided with the Suzuki of Dane Rene Rasmussen and the GSXR clip-on caught in the HRC Honda fairing and ripped off a radiator hose. The collision put Burnett up the slip road at the first corner. He rejoined the race but was forced to retire when the engine overheated after three laps.

Rymer led for the first twelve laps chased by the red Ducati of Monti, Pirovano on the Belgarda Yamaha and the HRC Hondas of Mertens and Merkel. Pirovano briefly held second until he slid off without injury with four laps completed.

Roche, on another Ducati, made his customary bad start but was forcing his way through the pack, spurred on by his legion of partisan fans. Also on a charge were the Bimota pair, Mike Baldwin and Giancarlo Falappa. At the mid point, they had latched onto the leading group and held second and fourth places respectively.

After three laps of shadowing Rymer, Baldwin used the superior top speed of the Bimota to good effect and drafted by Rymer into the lead. Less than two seconds behind a chasing group in third of Mertens, Falappa, Monti, Roche, Merkel, Coq and Tardozzi on another Bimota, the privately entered Speed Shadows example, were all battling among themselves whilst also trying to get on terms with the leaders.

Mertens by now had forged ahead and broke the tow between himself and the pursuing pack and eased by Rymer and then Baldwin to take the lead two laps from the end.

Baldwin was so intent on re-taking the lead, he failed to notice the booming Ducati of Roche riding shotgun and lost second place on the Mistral straight much to the delight of the small but vociferous crowd.

They willed Roche on to greater things but he was unable to catch the flying Mertens and even lost second to the determined Baldwin at the flag. Baldwin recorded the fastest lap in the race at 1m 21.90 on lap 11.

Rymer just managed to hold off repeated efforts by Falappa for fourth despite fighting some lurid slides in the late stages as his rear Dunlop had gone off.

But the victory went to Mertens who said, "I thought I could win that one. For sure it was a very tough race. There was a little problem with the tyres in the heat but I think we all had those problems. Baldwin rode very well and we had a good battle but I always thought that I could get by and win. I think the second race will be even harder.''

***RACE 2***

The Belgian proved correct. the weather was even hotter in the second race and so was the action on the track with Monti, Merkel and Rymer all getting scorching starts.

Once again, Monti's teammate, Roche, made a sluggish start and was eleventh as Monti led the field through for the first time with Merkel, Baldwin, Rymer, Mertens and Falappa forming a high speed freight train at the front of the pack.

Pirovano was already out with his OWO1 misfiring and Burnett also retired. He had elected to use his spare bike but it quickly lost power a dropped him out of the action.

At half distance, both Falappa and Baldwin had gotten the better of Merkel's Honda and were locked into their own private battle for the lead and this time Falappa had posted the fastest race lap at 1m 21.89, fractionally faster than Baldwin's race one effort. Mertens though was sitting ominously in fourth, looking to repeat his first race victory.

In fifth was hard riding Roche followed by a fading Monti, Anders Andersson, Rymer and Tardozzi. Up front Baldwin survived a hectic moment whilst trying to catch the flying Falappa. The rear end of the Bimota snapped sideways and Baldwin was pitched out of the bubble but somehow managed to hold onto the bars and regain control

''Jeez, that mother nearly threw me off,'' exclaimed Baldwin. ''I thought I was gonna hit the dirt for sure. I don't know how I got back on, but if anyone's got it on film, I'd sure like to see it.''

Baldwin, who faded to a fifth place finish behind Merkel after the incident, was not the only one having problems. Rymer was trying his hardest to stay in contact with the leaders and was pushing his tyres beyond their limits. After a few hairy moments, the front end gave way and he was pitched off into the dirt, luckily without injury with only four laps to run.

Up front Roche was again fighting to get back on terms but once again ran out of laps in his bid to catch Falappa and Mertens. He got Mertens but would have needed a couple of extra laps to pass the winning Bimota.

Mertens admitted thinking about retaking second place from Roche but opted for a

'safe' third and bagging points rather than risking a daring last lap pass.

At the flag, Falappa won by less that four tenths of a second from Roche and no doubt the crowd went away with the sight of the three leading men approaching the final turn in each others wheeltracks. A sight they will no doubt remember for a long time, a sight which has become a hallmark of World Superbike racing.

*Rymer chases Baldwin but the British rider later crashed heavily after pushing his Dunlops beyond their limits.*

# Qualifiers

# Round 6

# Paul Ricard, France

## 30 July

# Race Data

| POS | RIDER/NATIONALITY | ENTRANT/BIKE | TIME |
|---|---|---|---|
| 1. | Raymond Roche (F) | Squadra Corse Ducati 851S | 1m 21.83 |
| 2. | Fred Merkel (USA) | RCM/Rumi Honda RC30 | 1m 21.98 |
| 3. | Mike Baldwin (USA) | Bimota YB4EIR | 1m 22.16 |
| 4. | Baldassarre Monti (I) | Squadra Corse Ducati 851S | 1m 22.22 |
| 5. | Morris Coq (F) | Yamaha FZR750R | 1m 22.45 |
| 6. | Pierre Bolle (F) | Kawasaki ZXR750R | 1m 22.55 |
| 7. | Giancarlo Falappa (I) | Bimota Spa YB4EI | 1m 22.55 |
| 8. | Stephane Mertens (B) | Team Mediares Honda RC30 | 1m 22.58 |
| 9. | Terry Rymer (GB) | Loctite Yamaha OWO1 | 1m 22.89 |
| 10. | Jean Yves Mounier (F) | Yamaha FZR750R | 1m 22.94 |
| 11. | Jari Suhonen (SF) | Arwidson Yamaha FZR750R | 1m 23.00 |
| 12. | Davide Tardozzi (I) | Speed Shadows Bimota YB4EI | 1m 23.12 |
| 13. | Roger Burnett (GB) | HRC Honda RC30 | 1m 23.17 |
| 14. | Anders Andersson (S) | Yamaha Sweden FZR750R | 1m 23.24 |
| 15. | Christophe Bouheben (F) | Kawasaki France ZXR750 | 1m 23.52 |
| 16. | Fabrizio Pirovano (I) | Belgarda Yamaha FZR750R | 1m 23.55 |
| 17. | Rene Rasmussen (DK) | Pepsi Suzuki GSXR750R | 1m 23.75 |
| 18. | Paul Iddon (GB) | Pee Jay Yamaha FZR750R | 1m 23.80 |
| 19. | Eric Delcamp (F) | Yamaha FZR750R | 1m 23.95 |
| 20. | Andre Lussiana (F) | Honda RC30 | 1m 23.99 |
| 21. | Didier Hamdi (F) | Yamaha FZR750R | 1m 24.05 |
| 22. | Wolfgang von Muralt (CH) | Honda RC30 | 1m 24.05 |
| 23. | Rene Delaby (B) | Honda RC30 | 1m 24.08 |
| 24. | Bodo Schmidt (D) | Yamaha FZR750R | 1m 24.11 |
| 25. | Toni Rohrer (CH) | Honda RC30 | 1m 24.17 |
| 26. | Peter Rubatto (D) | Hein Gericke Bimota YB4EI | 1m 24.17 |
| 27. | Florian Ferracci (F) | Ducati 888 | 1m 24.20 |
| 28. | Patrick Igoa (F) | Kawasaki France ZXR750 | 1m 24.20 |
| 29. | Edwin Weibel (CH) | Kawasaki ZXR750 | 1m 24.22 |
| 30. | Andy McGladdery (GB) | Francis Neill Honda RC30 | 1m 24.44 |
| 31. | Jean Michel Mattioli (F) | Honda VFR750R | 1m 24.59 |
| 32. | Antonio Garcia Moreno (E) | Honda RC30 | 1m 24.64 |
| 33. | Mauro Ricci (I) | Ducati 851 | 1m 24.73 |
| 34. | Jabier Arenas (E) | Honda RC30 | 1m 24.85 |
| 35. | Piergiorgio Bontempi (I) | RCM/Rumi Honda RC30 | 1m 24.85 |
| 36. | Mile Pajic (NL) | Kawasaki ZXR750 | 1m 24.91 |
| 37. | Arnad Harmati (H) | Honda VFR750 | 1m 25.11 |
| 38. | Jeffrey de Vries (NL) | Yamaha FZR750R | 1m 25.14 |
| 39. | Emmanuel Lentaigne (F) | Kawasaki ZXR750R | 1m 25.14 |
| 40. | Sven Seidel (D) | Suzuki GSXR750R | 1m 25.18 |
| 41. | Jean Louis Tranois (F) | Suzuki GSXR750R | 1m 25.38 |
| 42. | Vittorio Scatola (I) | Honda RC30 | 1m 25.38 |
| 43. | Roberto Biagioli (I) | Yamaha FZR750R | 1m 25.45 |
| 44. | Michael Lundwall (S) | Kawasaki ZXR750R | 1m 25.62 |
| 45. | Giorgio Cantalupo (I) | Ducati 851 | 1m 25.70 |
| 46. | Marino Fabbri (I) | RB Moto Kawasaki ZXR750 | 1m 25.93 |

# Results Race 1

| POS | RIDER | LAPS | TIME | MPH | KPH |
|---|---|---|---|---|---|
| 1. | Mertens | 27 | 37m 28.03 | 88.61 | 142.69 |
| 2. | Baldwin | 27 | 37m 28.91 | 88.57 | 142.63 |
| 3. | Roche | 27 | 37m 29.11 | 88.56 | 142.62 |
| 4. | Rymer | 27 | 37m 32.48 | 88.43 | 142.40 |
| 5. | Falappa | 27 | 37m 38.27 | 88.21 | 142.04 |
| 6. | Mounier | 27 | 37m 38.63 | 88.19 | 142.02 |
| 7. | Monti | 27 | 37m 39.01 | 88.18 | 141.99 |
| 8. | Merkel | 27 | 37m 48.24 | 87.82 | 141.41 |
| 9. | Tardozzi | 27 | 37m 49.75 | 87.76 | 141.32 |
| 10. | Andersson | 27 | 38m 00.56 | 87.34 | 140.65 |
| 11. | Bolle | 27 | 38m 03.42 | 87.23 | 140.47 |
| 12. | Rasumussen | 27 | 38m 04.84 | 87.18 | 140.39 |
| 13. | de Vries | 27 | 38m 05.14 | 87.17 | 140.37 |
| 14. | Coq | 27 | 38m 06.53 | 87.12 | 140.28 |
| 15. | Hamdi | 27 | 38m 06.84 | 87.10 | 140.26 |
| 16. | Iddon | 27 | 38m 08.10 | 87.06 | 140.19 |
| 17. | Delcamp | 27 | 38m 16.24 | 86.74 | 139.69 |
| 18. | Weibel | 27 | 38m 21.42 | 86.55 | 139.38 |
| 19. | Bouheben | 27 | 38m 23.93 | 86.46 | 139.22 |
| 20. | Lussiana | 27 | 38m 24.43 | 86.44 | 139.19 |
| 21. | Bontempi | 27 | 38m 41.39 | 85.81 | 138.18 |
| 22. | Pajic | 27 | 38m 41.56 | 85.80 | 138.17 |
| 23. | Ricci | 27 | 38m 43.96 | 85.71 | 138.02 |
| 24. | Schmidt | 27 | 38m 46.41 | 85.62 | 137.88 |
| 25. | Arenas | 27 | 38m 46.54 | 85.61 | 137.87 |
| 26. | Lentaigne | 26 | 38m 00.29 | 84.12 | 135.46 |

**NOT CLASSIFIED**

Suhonen 25, Ferracci 21, Scatola 10, McGladdery 9, Moreno 8, Fabbri 7, Seidel 5, Pirovano 4, Burnett 3, Rubatto 2

| | |
|---|---|
| **Race Data:** | 2 x 28 laps of 2.04 mile circuit |
| **Total Distance:** | 114.24 miles |
| **Weather:** | Hot/sunny |
| **Fastest Lap:** | Giancarlo Falappa (race two) 1m 21.895s, 90.09 mph, 145.073 kph |
| **Existing Lap Record:** | Establishes new class record |

# Results Race 2

| POS | RIDER | LAPS | TIME | MPH | KPH |
|---|---|---|---|---|---|
| 1. | Falappa | 28 | 38m 44.40 | 88.87 | 143.11 |
| 2. | Roche | 28 | 38m 44.73 | 88.86 | 143.09 |
| 3. | Mertens | 28 | 38m 44.87 | 88.85 | 143.08 |
| 4. | Merkel | 28 | 38m 51.20 | 88.61 | 142.69 |
| 5. | Baldwin | 28 | 39m 00.37 | 88.26 | 142.13 |
| 5. | Monti | 28 | 39m 00.64 | 88.25 | 142.12 |
| 6. | Andersson | 28 | 39m 03.07 | 88.16 | 141.97 |
| 7. | Tardozzi | 28 | 39m 04.89 | 88.09 | 141.86 |
| 8. | Bolle | 28 | 39m 06.42 | 88.04 | 141.77 |
| 9. | Iddon | 28 | 39m 22.12 | 87.45 | 140.82 |
| 10. | Coq | 28 | 39m 23.21 | 87.41 | 140.76 |
| 11. | Rasmussen | 28 | 39m 30.46 | 87.14 | 140.33 |
| 12. | Mounier | 28 | 39m 32.19 | 87.08 | 140.23 |
| 13. | Weibel | 28 | 39m 34.29 | 87.00 | 140.10 |
| 14. | Delcamp | 28 | 39m 36.17 | 86.93 | 139.99 |
| 15. | Lussiana | 28 | 39m 55.58 | 86.23 | 138.86 |
| 16. | Garcia Moreno | 28 | 40m 01.27 | 86.03 | 138.53 |
| 17. | McGladdery | 28 | 40m 03.71 | 85.94 | 138.39 |
| 18. | Ricci | 28 | 40m 05.09 | 85.89 | 138.31 |
| 19. | Pajic | 28 | 40m 05.73 | 85.87 | 138.27 |
| 20. | Arenas | 28 | 40m 06.46 | 85.84 | 138.23 |
| 21. | Ferracci | 28 | 40m 10.14 | 85.71 | 138.02 |

**NOT CLASSIFIED**

Rymer 24, Rubatto 17, Bontempi 17, Burnett 10, Lentaigne 9, Fabbri 8, Pirovano 4, Scatola 4, Schmidt 3

# Championship Positions

**1.** Merkel 167, **2.** Mertens 147, **3.** Pirovano 133, **4.** Roche 129, **5.** Andersson 99, **6.** Rymer 98, **7.** Falappa 94, **8.** Monti 71, **9.** Baldwin 55, **10.** McMurter, Suhonen 38, **12.** Burnett 32, **13.** Igoa 31, **14.** McElnea, Vieira, Gschwender 28, **17.** Tardozzi 27, **18.** Broccoli, Rasmussen 26, **20.** Campbell 24, **21.** Douglas 20, **22.** Biliotti, Morrison, Rubatto 17, **25.** Goodfellow 16, **26.** Hofmann, Iddon, Bolle 14, **29.** Mounier, Hislop 13, **31.** Fogarty 12, **32.** Crine, Phillips, Crevier 11, **35.** Mercier, Delaby, Coq 7, **38.** Fabbri 6, **39.** McGladdery 5, **40.** von Muralt, Ricci, de Vries 3, **43.** Moineau, Weibel 2, **45.** Bouheben, Russell, Hamdi, Delcamp 1

# Round 7

# Sugo, Japan

## 27 August

*The orange 'Go' sign shines like a beacon in the dismal Sugo weather.*

*Aussie Mike Dowson nurses an injured throttle hand with an ice pack after a heavy tumble in training.*

Texan Doug Polen gave Suzuki their first World Superbike victory of the season when he won the first race on his Yoshimura prepared GSXR750R at a rain lashed Sugo. The second leg was won by OWO1 Yamaha mounted Aussie ace Michael Dowson amidst typhoon like conditions. Dowson had finished second in the first leg, seven seconds adrift but was the rain master in the second, winning by a huge 24 second margin.

Another Aussie, Rob Phillis, gave Kawasaki something to cheer about in the first leg. He became the first rider to get a ZXR750 into a rostrum placing in 1989 - and this was his first World Superbike event of the season!

None of these riders figured in the overall championship picture. Giancarlo Falappa, having gone 4-3 in the two races, was by far the best European performer while points leader and reigning champion Fred Merkel saw his overall advantage slip to only eight points over Belgian Stephane Mertens.

Merkel's wet weather Pirelli's failed to reach working temperature in the first leg and he slipped to 16th, and out of the points and was only 13th in the second race. Mertens meantime scored a pair of steady eighth places to continue his play for the title.

*In spite of his injured hand, Dowson was the rain master of the second leg on his factory supplied FZR750R OWO1 Yamaha.*

***ENTRY/QUALIFYING***

Weather conditions early in practice were good and no less than eleven riders were inside the old World Superbike lap record of 1m 36.73 set by Edwin Weibel in 1988. Doug Polen, based in Japan with the Sietto GP-1 Yoshimura team, grabbed pole with a scorching 1m 35.12 but was only 0.06 seconds quicker than Fred Merkel. Four different marques lined up on the front row Raymond Roche on the ZDM851 Ducati and Aussie Kawasaki rider, Rob Phillis completing the line up with 1m 35.72 and 1m 35.42 respectively.

As ever, qualifying was extremely competitive and several riders took heavy tumbles in search of a quick time. Bimota teamster Giancarlo Falappa crashed heavily and badly damaged his left knee. It was thought he had torn some ligaments and his cartilage but the gritty Italian decided to take part in the race anyway and had his seat padded so he didn't have to bend the injured limb so far to reach the foot peg.

Another faller was Michael Dowson who slid off his Australian entered, Marlboro Yamaha Dealer Team OWO1 Yamaha at the chicane and injured his right hand. By the time he had trudged back to the pits, his hand had swollen like a balloon and it was only continuous use of an ice pack that enabled him to get his racing gloves back on.

The most unfortunate victim of all the

SUZUKI
Kyoseki
SUZUKI
MICHELIN
Arai
RK

*American Doug Polen gave Suzuki their first World Superbike win of the season when he sped to a seven second victory in the first leg at Sugo on his Sietto GP1 Yoshimura GSXR750R.*

practice tumbles was Finnish rider Jari Suhonen (Arwidson Yamaha) who re-broke the ankle he had damaged in the previous French round.

***RACE 1***

The typhoon which had been threatening the coast of Japan the previous day, finally started to make its presence felt on race day. Fog and heavy clouds hung around the circuit and by the time the first race gridded, it was already drizzling steadily.

A few of the riders gambled on intermediates, notably Ducati's Baldassare Monti, while the remainder settled for full wets. Within two or three laps, Monti and the others on inters had realised their big mistake. The track was awash as the rain teemed down.

Phillis had made a superb start and the green machine led the field down the hill to the first corner followed by Polen, Dowson, Merkel and Super Angel YZR750 Yamaha mounted, Andrew Stroud.

By the end of the first lap, Phillis still led but Merkel was out of contention as he slithered around on his unsuitable Pirelli wets.

After three laps, Polen assumed control after deciding his close shadows, Dowson

*The Superbike regulars had a tough time in Japan. Here, Pirovano splashes through a chicane in front of Falappa and Campbell.*

and Stroud had gotten too close for comfort. As he slipped by Phillis for the lead, Dowson and Stroud responded in wildy differing ways.

Dowson overtook Phillis as well and got into the Suzuki's wheel tracks while Stroud dumped it in a big way on the fourth lap.

Dowson actually passed Polen and tried to make a break but pushing hard, lost the front in a big way and went grasstracking in an effort to kept upright. He just held on but immediately lost his lead.

Undeterred he got his head down but a lap later astonished the crowd with a superb piece of rodeo riding to get out of yet another lurid moment. Exiting turn two he got completely out of shape and was thrown out of the saddle up into the air. Somehow he gripped hold of the bars but when he bounced back down into the saddle his right leg got caught under the bike and was almost run over by his rear wheel.

Explaining the incident, he said, "I was off the bike and I could feel the rear tyre hitting the back of my leg. I though I was gonna run myself over. Suddenly she straightened out and I was able to jump back on board but by then Doug (Polen) was out of sight. I though I'd better take it a bit easier then and settle for second."

The pattern of the race was set. Polen was in command up front and pulled away from Dowson to win by some seven seconds and displaying the form that had taken him to the leadership of the All Japan Formula One Championship. In second and third were Aussies Dowson and Phillis circulating quickly but safely.

Meantime, Falappa was moving through the field in a controlled yet determined manner. Because of the deteriorating conditions and his injured knee he was having to control his normally aggressive behaviour and ride more smoothly picking up places lap by lap.

A little further back, Pirovano and Mertens were involved in an interesting tussle. The Belgian was unable to find a way past the diminutive Italian without risking a crash so as in France, he settle for points and slotted into eighth place with his pit crew keeping him constantly informed about the lack of progress of his chief rival, Merkel.

As the race ran its wet course, Polen completed his lonely race and admitted it was difficult for him to maintain concentration with no one challenging him. He said, "I sure am pleased for Suzuki, and of course, Yoshimura. I thought I could win that one right from the beginning. It was a bit slippery out there and I had a few slides, but it was all controlled stuff.

"I'd tried some special Michelin wet tyres in practice but decided to use the standard ones and they worked really fine. Towards the end of the race I tried some new lines through the turns but I kept it smooth. The last thing I wanted to do was fall off so close to the chequered flag."

### *RACE 2*

The drizzle had turned to lashing rain accompanied by high winds as the grid assembled for race two. In the murky conditions the 'GO' light shone out like a brilliant green beacon as the field slithered and wheelspun away off the line. Once again it was the green meanie of Phillis out in front chased by Polen, Dowson, Team Kawasaki's Takahiro Sohwa Aaron Slight and Kenichiro Iwahashi on the BVD Honda.

Within three laps Dowson had got in front and immediately started extending his lead despite the diabolical conditions. As the rain's intensity increased, there were rivers of water flowing across the track making the likelihood of aquaplaning a serious threat.

Falappa had gotten off the line really well this time but was bumped off course at the end of the very first turn which left him much work to do. Showing scant regard for the conditions, he fought his way through the field until he was on the tail of race one winner, Polen.

"At the end of the first lap I was not very happy," grimaced Giancarlo. "I was a long way down the field and was having trouble with the water temperature gauge. The motor took a long time to reach the right temperature and then I wanted to get as far forward in the race as I could. I am a very aggressive rider in the wet and in the dry and every time I enter a race, I want to win. I passed many people and then I was behind Polen. I knew I could overtake him in the chicane, but waited until the end of the straight so that all the people could see!"

At the front, Dowson was in command and no one could get near him as he splashed his way to victory. Second some twenty seconds adrift was local rider Iwahashi on an RC30 with Falappa crowning his superb days racing with an excellent third place.

At the flag, Polen was just two tenths of a second adrift of the Bimota star. Of the championship aspirants, Mertens grabbed another eighth place while Merkel could only salvage twelfth place points. Pirovano had not fared at all well in Japan. Following his dnf in race one, he crashed out in the second leg after 15 laps when the bike slipped from under him at the Esses.

So Dowson received the winners garland and once again showed that Australia is a tremendous breeding ground for top league fourstroke racers, maybe even better than America can currently lay claim to.

After the rostrum ceremonies, Dowson said, "Yeah I really enjoyed that. I had no problems at all, even my hand didn't hurt. The rain was terrible, there were rivers running across the track in places and the wind was really gusty. It also seemed to come from different directions, especially after turn six. I took it real carefully out there and I had great signals from my pit crew so I knew what was going on with all the guys behind me.

"The bike worked fine, the tyres worked good and I just hung on in there and let the other guys do the worrying. The race felt it was going on forever and it was getting kind of murky in the later stages but I was in front and I didn't care. I was a bit surprised no one got anywhere near me though."

# Qualifiers

# Round 7

# Sugo, Japan

## 27 August

# Race Data

| POS | RIDER/NATIONALITY | ENTRANT/BIKE | TIME |
|---|---|---|---|
| 1. | Doug Polen (USA) | Yoshimura Suzuki GSXR750R | 1m 35.12 |
| 2. | Fred Merkel (USA) | Team Rumi Honda RC30 | 1m 35.18 |
| 3. | Rob Phillis (AUS) | Team Kawasaki Aus/ZXR750 | 1m 35.42 |
| 4. | Raymond Roche (F) | Squadra Corse Ducati | 1m 35.72 |
| 5. | Takahiro Sohwa (J) | KRT Kawasaki ZXR750 | 1m 35.87 |
| 6. | Grant Hodson (AUS) | Hayashi/Yamaha FZR750R | 1m 36.11 |
| 7. | Michael Dowson (AUS) | Yamaha Australia FZR750R | 1m 36.22 |
| 8. | Shingo Katoh (J) | YA Racing Yamaha FZR750R | 1m 36.33 |
| 9. | Stephane Mertens (B) | Total/Bel Ray Honda | 1m 36.38 |
| 10. | Shoichi Tsukamoto (J) | KRT Kawasaki ZXR750 | 1m 36.62 |
| 11. | Tatsuro Arata (J) | Over Racing/Yamaha FZR750R | 1m 36.71 |
| 12. | Davide Tardozzi (I) | Speed Shadows /Bimota | 1m 36.77 |
| 13. | Malcolm Campbell (AUS) | Citizen Honda VFR750R | 1m 36.79 |
| 14. | Aaron Slight (NZ) | Kawasaki Australia ZXR750 | 1m 36.88 |
| 15. | Kenichiro Iwahashi(J) | Yamamoto Honda VFR750R | 1m 36.89 |
| 16. | Baldassarre Monti (I) | Squadra Corse Ducati | 1m 37.20 |
| 17. | Rueben McMurter (CDN) | Shoei Honda VFR750R | 1m 37.26 |
| 18. | Koh Takeuchi (J) | Hayashi Yamaha FZR750R | 1m 37.27 |
| 19. | Gary Goodfellow (CDN) | Don Knit Yoshimura Suzuki | 1m 37.41 |
| 20. | Toshiyuki Takeda (J) | Honda VFR750R | 1m 37.48 |
| 21. | Terry Rymer (GB) | Loctite Yamaha FZR750R | 1m 37.55 |
| 22. | Anders Andersson (S) | Yamaha Sweden FZR750R | 1m 37.59 |
| 23. | Patrick Igoa (F) | Team Kawasaki France | 1m 37.59 |
| 24. | Fabrizio Pirovano (I) | Belgarda Yamaha FZR750R | 1m 37.77 |
| 25. | Giancarlo Falappa (I) | Bimota Spa YB4EI | 1m 37.79 |
| 26. | Masashi Satoh (J) | Yamaha FZR750R | 1m 37.86 |
| 27. | Naoto Abe (J) | Honda VFR750R | 1m 38.01 |
| 28. | Kohsei Yamamoto (J) | Honda VFR750 | 1m 38.15 |
| 29. | Mike Baldwin (USA) | Bimota Spa YB4EI | 1m 38.16 |
| 30. | Andrew Stroud (NZ) | Yamaha FZR750R | 1m 38.38 |
| 31. | Jari Suhonen (SF) | Arwidson Yamaha FZR750R | 1m 38.41 |
| 32. | Roger Burnett (GB) | HRC Honda RC30 | 1m 38.42 |
| 33. | Yukiya Ohshima (J) | Yoshimura Suzuki GSXR750R | 1m 38.45 |
| 34. | Wayne Clarke (AUS) | Yamaha FZR750R | 1m 38.51 |
| 35. | Christophe Bouheben (F) | Team Kawasaki France | 1m 38.60 |
| 36. | Kohichi Kobayashi (J) | Blue Helmet Honda VFR750R | 1m 38.60 |
| 37. | Takayoshi Yamamoto (J) | Moto Sports Yamaha FZR750R | 1m 38.62 |
| 38. | Shigemasa Miwa (J) | Blue Helmet Honda VFR750R | 1m 38.74 |
| 39. | Toshiharu Kaneko (J) | Yamaha FZR750R | 1m 38.78 |
| 40. | Scott Doohan (AUS) | Bimota YB4 | 1m 38.91 |

**Race Data:** 2 x 25 laps of 2.32 mile circuit
**Total Distance :** 116 miles
**Weather:** Rain
**Fastest Lap:** Doug Polen (race one) 1m 50.625
**Existing Lap Record:** Edwin Weibel 1m 36.73 (August 1988)

# Results Race 1

| POS | RIDER | LAPS | TIME | MPH | KPH |
|---|---|---|---|---|---|
| 1. | Polen | 25 | 47m 06.50 | | |
| 2. | Dowson | 25 | 47m 13.82 | | |
| 3. | Phillis | 25 | 47m 18.42 | | |
| 4. | Falappa | 25 | 47m 20.88 | | |
| 5. | Arata | 25 | 47m 26.41 | | |
| 6. | Iwahashi | 25 | 47m 30.47 | | |
| 7. | Pirovano | 25 | 47m 35.53 | | |
| 8. | Mertens | 25 | 47m 35.74 | | |
| 9. | Slight | 25 | 47m 39.88 | | |
| 10. | Abe | 25 | 48m 02.6 | | |
| 11. | Tsukamoto | 25 | 48m 08.83 | | |
| 12. | Tardozzi | 25 | 48m 10.47 | | |
| 13. | Roche | 25 | 48m 12.47 | | |
| 14. | Doohan | 25 | 48m 20.34 | | |
| 15. | Campbell | 25 | 48m 23.17 | | |
| 16. | Merkel | 25 | 48m 24.97 | | |
| 17. | Burnett | 25 | 48m 27.75 | | |
| 18. | T. Yamamoto | 25 | 48m 43.21 | | |
| 19. | Kobayashi | 25 | 48m 44.38 | | |
| 20. | Takeuchi | 25 | 48m 48.95 | | |
| 21. | Clarke | 24 | 47m 12.84 | | |
| 22. | Delcamp | 24 | 47m 14.59 | | |
| 23. | McMurter | 24 | 47m 57.05 | | |
| 24. | Miwa | 24 | 48m 07.16 | | |
| 25. | Crafar | 23 | 47m 17.58 | | |
| 26. | Kaneko | 22 | 47m 11.45 | | |
| 27. | Hodson | 20 | | | |

**NOT CLASSIFIED**

K. Yamamoto 14 laps, Monti 11, Igoa 10, Rymer 7, Katoh 4, Stroud 3,Bouheben 2,Sohwa 1, Satoh 1, Baldwin 1, Andersson 1

# Results Race 2

| POS | RIDER | LAPS | TIME | MPH | KPH |
|---|---|---|---|---|---|
| 1. | Dowson | 25 | 47m 45.22 | | |
| 2. | Iwahashi | 25 | 48m 09.91 | | |
| 3. | Falappa | 25 | 48m 15.56 | | |
| 4. | Polen | 25 | 48m 15.74 | | |
| 5. | Sohwa | 25 | 48m 45.66 | | |
| 6. | Phillis | 25 | 48m 59.03 | | |
| 7. | Slight | 25 | 49m 04.47 | | |
| 8. | Mertens | 25 | 49m 07.20 | | |
| 9. | Arata | 25 | 49m 07.93 | | |
| 10. | Campbell | 25 | 49m 08.12 | | |
| 11. | Tsukamoto | 25 | 49m 25.85 | | |
| 12. | Merkel | 25 | 49m 29.89 | | |
| 13. | Rymer | 25 | 49m 32.58 | | |
| 14. | Kobayashi | 25 | 49m 43.91 | | |
| 15. | Takeuchi | 25 | 49m 46.49 | | |
| 16. | Abe | 24 | 47m 50.84 | | |
| 17. | T. Yamamoto | 24 | 47m 52.86 | | |
| 18 | Roche | 24 | 47m 59.81 | | |
| 19. | Satoh | 24 | 48m 12.70 | | |
| 20. | Baldwin | 24 | 48m 13.08 | | |
| 21. | Stroud | 24 | 48m 49.51 | | |
| 22. | Doohan | 24 | 49m 25.38 | | |
| 23. | Miwa | 24 | 49m 27.66 | | |

**NOT CLASSIFIED**

Katoh 22 laps, Delcamp 22, Kaneko 22, Crafar 21, Hodson 19, Pirovano 15, Bouheben 14, Monti 14, McMurter 11, Igoa 6, Andersson 5, K. Yamamoto 3, Tardozzi 1, Burnett 1, Clarke 0

***No further information supplied by race organisers.***

# Championship Positions

**1.** Merkel 171, **2.** Mertens 163, **3.** Pirovano 142, **4.** Roche 132, **5.** Falappa 122, **6.** Rymer 101, **7.** Andersson 99, **8.** Monti 71, **9.** Baldwin 55, **10** McMurter, Suhohen 38, **12.** Dowson 37, **13.** Polen 33, **14.** Burnett 32, **15.** Igoa, Tardozzi, Campbell 31, **18.** McElnea, Vieira, Gschwender 28, **21.** Broccoli, Iwahashi 27, **23.** Rasmussen 26, **24.** Phillis 25, **25.** Douglas 20, **26.** Arata 18, **27.** Biliotti, Morrison, Rubatto 17, **30.** Goodfellow, Slight 16, **32.** Hofmann , Iddon, Bolle 14, **35.** Mounier, Hislop 13, **37.** Fogarty 12, **38.** Crine, Phillips, Crevier, Sohwa 11, **42.** Tsukamoto 10, **43.** Mercier, Delaby, Coq 7, **46.** Fabbri, Abe 6, **48.** McGladdery 5, **49.** von Muralt, Ricci, de Vries 3, **52.** Moineau, Weibel, Doohan, Kobayashi 2, **56.** Bouheben, Russell, Hamdi, Delcamp, Takeuchi 1

*Anders Andersson was on the front row at Hockenheim with his Yamaha.*

# Round 8

# Hockenheim, West Germany

## 17 September

*Pirovano leads Bimota men Falappa and Baldwin through the stadium.*

*Jubilant Roche (centre) and Mertens (right) acknowledge the crowd while hard faced Falappa looks on.*

*Andersson (21) and Tardozzi (3) make the break.*

YAMAHA
2
6
38
MICHELIN

*Pirovano (2), Vieira (6), Hofmann (38) play to packed stands.* ***(Above)*** *Alex Vieira picked up points in another of his infrequent Superbike outings on the Honda France RC30. His best result at Hockenheim came in the first leg when he took third place.*

Hockenheim is a horsepower track, make no mistake. So it was no surprise to see Raymond Roche taking both races with the blindingly fast factory Ducati. It was the second time the French ace had achieved a double victory in a Superbike round. He also dominated the Brainerd, North America meeting - another high speed circuit.

The 40 point haul put him within 23 points of series leader Stephane Mertens who went 2-3 at Hockenheim while Californian Fred Merkel went three points down on the Belgian after a his second successive disappointing visit to the West German venue.

Claiming his HRC supplied Honda was down on speed compared to Roche's Ducati, something which made Mertens' Honda tuner Jean D'Hollander smile because he had spent a lot of time on the test bench ensuring his Honda would compete with the powerful vee twin, Merkel could only finish eighth and fourth in the respective races.

In the first leg Roche fought off a strong challenge from Mertens and Alex Vieira before a piece of luck helped him win the day. Then in the second leg he fended off a challenge from Giancarlo Falappa and Mertens before edging clear to a comfortable two second victory.

### *ENTRY/QUALIFYING*

Incredibly the entry topped the 100 mark for this eighth round of the championship but it was one man, Raymond Roche who dominated qualifying, speeding to a pole winning time of 2m 9.58s

Once again qualifying showed just how competitive Superbike racing has become.

*Raymond Roche had the right combination at Hockenheim. A lot of experience and a very fast motorcycle.*

Mobil 1
27

All five bikes on the front row of the grid were from different manufacturers.

After Roche came, Mertens (Honda) a mere 0.1s adrift, Italian Davide Tardozzi with a time of 2m 10.54s on the Speed Shadows Racing Bimota, Anders Andersson clocked a 2m 10.90s on the Yamaha Sweden bike and Frenchman Patrick Igoa completed the front row with a 2m 11.01 on the Godier Kawasaki.

Of that elite collection of motorcycles, the biggest surprise was that of Igoa on the Kawasaki. Until Hockenheim the bike had lacked speed but there was clearly no shortage of horsepower for the green meanie now.

Georges Godier, former Coupe d'Endurance champion and now team manager of the Kawasaki France entry explained, ''all season we have been waiting for some good bits from the factory in Japan. At last we've got them, but now it's too late. there's only a couple of races left.''

There was a surprise too for the Superbike regulars when local rider, Andreas Hofmann clocked a 2m 11.07s, the sixth fastest time. The Eckert Honda man had already impressed earlier in the year in an F1 race at Assen and was keen to play the spoiler role on his home circuit.

Falappa was quickest of the Bimotas in seventh (teammate Baldwin was only 0.77s slower but that dropped the American to 15th on the grid!) then came Merkel, posting a 2m 11.25. The Superbike champion had had an appalling time at the same circuit the previous season when he was hampered with a slow bike and failed to score. This season he was hoping the meeting would yield a better points return.

Ducati rider Baldassare Monti completed the second row with a 2m 11.62 while his team boss, Marco Lucchinelli having a rare racing outing since his appointment to a managerial role with Ducati, was on the third row with a 2m 11.67.

***RACE 1***

Roche failed to take full advantage of his pole position and managed to hold up Pirovano who was making his usual holeshot, this time from the third row.

As the bunch filed into the stadium at the end of the first lap they were greeted by a huge roar from the assembled masses. And no wonder! Hofmann was leading from Mertens, Vieira (who had only qualified 14th) and Roche.

These four had a small cushion before the main bunch led by Andersson, Pirovano, Merkel and Igoa appeared. Hofmann soon found the pace at the front too hot and was passed first by Mertens and then Roche and Vieira but the leading threesome became a duo as Mertens and Roche went head to head at the front.

The Ducati had the better grunt out of the turns, all important at Hockenheim where the long straights out in the country are punctuated by tight chicanes, but Mertens hung in there giving rise to speculation that he might be riding a tactical race and was simply waiting to pick his moment to pass

Roche was faring considerably better than his Ducati team manager Lucchinelli who had stopped at the end of the third lap with an electrical problem.

Nearing the final stages, Mertens indeed got by but an ambulance car left the pit lane in a hurry to reach the scene of a crash and pulled out in front of the two leaders. Mertens had his head down behind the screen and didn't see the car while Roche observed the bizarre action from behind the Belgian's Honda and was able to pick his line around the obstacle.

Mertens thankfully saw the car at the last minute as he sat up to brake but by then he had to take evasive action to avoid a collision and lost considerable time getting onto the fast straight leading out into the country.

Roche had the better line around the turn at the end of the start/finish straight and that was enough to give him the winning advantage. No matter what he tried, Mertens was unable to get back on terms to challenge the red Ducati.

Roche ran out the winner by just under a second while Vieira took a solid third, some six seconds adrift.

Pirovano was fourth, just beating Andersson's similar Yamaha to the line and Baldwin got sixth on the Bimota. Early leader Hofmann eventually wound up a creditable seventh, over two seconds ahead of a disappointed Merkel.

After the race Roche said, ''I made a bad start and had to do a lot of work to catch up. My Ducati is so fast here so it was not too much of a problem. I was behind Stephane at the end of the straight with about two or three laps to go and deciding where to pass him. I could see that he was tucked down behind the screen and that he did not see the ambulance suddenly come out onto the track. He had to slow down, but I was just able to go past without slowing at all. After that I had no trouble in winning.''

***RACE 2***

The second leg started in blazing sunshine and off to an equally blazing start was Anders Andersson who led into the first turn, chased by Falappa, Mertens, Merkel and Roche.

Even before the end of lap one, Roche had forced by everyone except new leader Mertens and was even threatening him as they entered the stadium for the first time.

On lap two the irrepressible Frenchman sped by into a lead he was never to lose but the race behind was far from over. The main fight was between Mertens and Falappa for second with the Italian charger using every inch of the track, and plenty of the dirt too, in his efforts to keep the Honda at bay, while the cool Belgian sat in there watching the antics and waiting for any possible error.

*Giancarlo Falappa came out on top after a hectic second race, second place battle with Stephane Mertens.*

The race developed into a series of battles. apart from the exciting Falappa/Mertens confrontation. Merkel, going much better this time around, was scraping furiously with Andersson while slightly further back, Pirovano, Baldwin and Vieira were involved in a tremendous dice.

Roche ran out an easy winner by a clear two seconds but the other places were only settled on the final lap. Mertens decided it was too dangerous to risk passing the erratic Falappa and settled for third.

Merkel got the better of Andersson when the likeable Swede's tyres went off in the late stages. Merkel is not a lover of the Hockenheim circuit so he was delighted with his fourth place which kept him within three points of series leader Mertens.

Further back Pirovano won the sixth place spot from Baldwin and Vieira but they finished line astern separated by less than half a second. Igoa was a distant ninth and Lucchinelli a lonesome tenth.

So Roche gained a maximum 40 point haul and moved back into contention for the title with a machine that looked at last to have found some meaningful reliability.

He was happy and looking forward to the remaining three rounds as was Mertens who now had the points lead, albeit by a slender three point margin. With the ever improving form of Roche and Mertens, Merkel left for Sicily a worried man. The championship was heading for a grandstand finish and it would take a brave man to predict the outcome.

# Qualifiers

# Round 8

# Hockenheim, West Germany

## 17 September

# Race Data

| POS | RIDER/NATIONALITY | ENTRANT/BIKE | TIME |
|---|---|---|---|
| 1. | Raymond Roche (F) | Squadra Corse Ducati | 2m 09.58 |
| 2. | Stephane Mertens (B) | Team Mediares Honda RC30 | 2m 09.68 |
| 3. | Davide Tardozzi (I) | Speed Shadows Bimota YB4EI | 2m 10.54 |
| 4. | Anders Andersson (S) | Yamaha Sweden FZR750R | 2m 10.90 |
| 5. | Patrick Igoa (F) | Kawasaki France ZXR750 | 2m 11.01 |
| 6. | Andreas Hofmann (D) | Honda RC30 | 2m 11.07 |
| 7. | Giancarlo Falappa (I) | Bimota Spa YB4EI | 2m 11.20 |
| 8. | Fred Merkel (USA) | RCM/Rumi Honda RC30 | 2m 11.25 |
| 9. | Baldassarre Monti (I) | Squadra Corse Ducati 851S | 2m 11.62 |
| 10. | Fabrizio Pirovano (I) | Belgarda Yamaha FZR750R | 2m 11.63 |
| 11. | Marco Lucchinelli (I) | Squadra Corse Ducati | 2m 11.67 |
| 12. | Michael Galinski (D) | Mitsui Yamaha FZR750R | 2m 11.73 |
| 13. | Peter Rubatto (D) | Hein Gericke Bimota YB4EI | 2m 11.73 |
| 14. | Alex Vieira (P) | Honda Motul RC30 | 2m 11.79 |
| 15. | Mike Baldwin (USA) | Bimota YB4EIR | 2m 11.99 |
| 16. | Ernst Gschwender (D) | Suzuki Deutschland GSXR750 | 2m 12.15 |
| 17. | Bodo Schmidt (D) | Yamaha FZR750R | 2m 12.31 |
| 18. | Jean Michel Mattioli (F) | Honda VFR750R | 2m 12.54 |
| 19. | Jari Suhonen (SF) | Arwidson Yamaha FZR750R | 2m 12.59 |
| 20. | Udo Mark (D) | Zweirad Gaeckle Yamaha | 2m 12.60 |
| 21. | Roger Burnett (GB) | Honda UK RC30 | 2m 12.64 |
| 22. | Paul Iddon (GB) | Pee Jay Yamaha FZR750R | 2m 12.78 |
| 23. | Graeme McGregor (AUS) | Honda RC30 | 2m 12.89 |
| 24. | Anton Gruschka (D) | Team Honda Shop RC30 | 2m 13.01 |
| 25. | Manfred Fischer (D) | Schuh Racing Honda RC30 | 2m 13.34 |
| 26. | Jeffry de Vries (NL) | Yamaha FZR750R | 2m 13.38 |
| 27. | Christian Monsch (CH) | Grisoni Racing Honda RC30 | 2m 13.83 |
| 28. | W. Moeckel (D) | Honda RC30 | 2m 13.85 |
| 29. | Edwin Weibel (CH) | Kawasaki ZXR750 | 2m 13.94 |
| 30. | Marino Fabbri (I) | RB Moto Kawasaki ZXR750 | 2m 14.00 |
| 31. | Herve Moineau (F) | Minolta Suzuki GSXR750 | 2m 14.00 |
| 32. | Jean Yves Mounier (F) | Yamaha FZR750R | 2m 14.04 |
| 33. | Wolfgang von Muralt (CH) | Honda RC30 | 2m 14.64 |
| 34. | Anton Heiler (D) | Yamaha Emonts FZR750R | 2m 15.21 |
| 35. | L. Fahr (D) | Thiede Power Honda RC30 | 2m 15.46 |
| 36. | M. Kaiser (D) | Yamaha FZR750R | 2m 16.07 |

**Race Data:** 2 x 14 laps of 4.21 mile circuit
**Total Distance:** 117.88 miles
**Weather:** Sunny
**Fastest Lap:** Raymond Roche (race two) 2m 08.189s, 117.92 mph. 189.888 kph
**Existing Lap Record:** Virginio Ferrari 2m 11.78s, 115.15 mph, 185.438 kph (May 1988)

## Results Race 1

| POS | RIDER | LAPS | TIME | MPH | KPH |
|---|---|---|---|---|---|
| 1. | Roche | 14 | 30m 30.94 | 116.04 | 186.85 |
| 2. | Mertens | 14 | 30m 31.91 | 115.97 | 186.75 |
| 3. | Vieira | 14 | 30m 37.84 | 115.60 | 186.15 |
| 4. | Pirovano | 14 | 30m 38.11 | 115.58 | 186.12 |
| 5. | Andersson | 14 | 30m 38.38 | 115.57 | 186.10 |
| 6. | Baldwin | 14 | 30m 41.13 | 115.39 | 185.82 |
| 7. | Hofmann | 14 | 30m 44.31 | 115.19 | 185.50 |
| 8. | Merkel | 14 | 30m 46.85 | 115.04 | 185.24 |
| 9. | Igoa | 14 | 30m 59.89 | 114.23 | 183.94 |
| 10 | Suhonen | 14 | 31m 05.27 | 113.90 | 183.41 |
| 11. | Schmidt | 14 | 31m 05.84 | 113.86 | 183.36 |
| 12 | Burnett | 14 | 31m 15.85 | 113.26 | 182.38 |
| 13. | Weibel | 14 | 31m 16.10 | 113.24 | 182.35 |
| 14. | Mounier | 14 | 31m 16.32 | 113.23 | 182.33 |
| 15. | McGregor | 14 | 31m 16.52 | 113.22 | 182.31 |
| 16. | Galinski | 14 | 31m 20.62 | 112.97 | 181.92 |
| 17. | Moeckel | 14 | 31m 21.30 | 112.93 | 181.85 |
| 18. | Mark | 14 | 31m 22.10 | 112.88 | 181.77 |
| 19. | von Muralt | 14 | 31m 23.47 | 112.80 | 181.64 |
| 20. | Fischer | 14 | 31m 41.44 | 111.73 | 179.92 |
| 21. | Heiler | 14 | 31m 58.18 | 110.76 | 178.35 |
| 22. | Franz | 14 | 32m 07.71 | 110.21 | 177.47 |
| 23. | Krummenacker | 14 | 32m 11.81 | 109.98 | 177.10 |
| 24. | Fahr | 14 | 32m 12.52 | 109.94 | 177.03 |

**NOT CLASSIFIED**

Iddon 14 laps, Gruschka 10, Bouheben 10, de Vries 10, Gschwender 7, Kaiser 7, Moineau 6, Falappa 5, Mattioli 4, Tardozzi 4, Lucchinelli 3, Monsch 2, Monti 2, Cantalupo 2, Rubatto 1

## Results Race 2

| POS | RIDER | LAPS | TIME | MPH | KPH |
|---|---|---|---|---|---|
| 1. | Roche | 14 | 30m 32.08 | 116.60 | 187.76 |
| 2. | Falappa | 14 | 30m 24.40 | 116.45 | 187.52 |
| 3. | Mertens | 14 | 30m 24.97 | 116.41 | 187.46 |
| 4. | Merkel | 14 | 30m 33.59 | 115.87 | 186.58 |
| 5. | Andersson | 14 | 30m 34.74 | 115.79 | 186.47 |
| 6. | Pirovano | 14 | 30m 41.63 | 115.36 | 185.77 |
| 7. | Baldwin | 14 | 30m 41.89 | 115.35 | 185.74 |
| 8. | Vieira | 14 | 30m 42.12 | 115.33 | 185.72 |
| 9. | Igoa | 14 | 30m 47.24 | 115.01 | 185.20 |
| 10. | Lucchinelli | 14 | 30m 52.47 | 114.69 | 184.68 |
| 11. | Schmidt | 14 | 30m 55.12 | 114.52 | 184.42 |
| 12. | Hofmann | 14 | 30m 55.31 | 114.51 | 184.40 |
| 13. | Mark | 14 | 30m 55.60 | 114.49 | 184.37 |
| 14. | Suhonen | 14 | 31m 10.90 | 113.56 | 182.86 |
| 15. | Bouheben | 14 | 31m 16.54 | 113.22 | 182.31 |
| 16. | Moeckel | 14 | 31m 17.21 | 113.17 | 182.25 |
| 17. | de Vries | 14 | 31m 17.38 | 113.16 | 182.23 |
| 18. | McGregor | 14 | 31m 17.76 | 113.14 | 182.19 |
| 19. | Galinski | 14 | 31m 20.00 | 113.01 | 181.98 |
| 20. | von Muralt | 14 | 31m 26.62 | 112.61 | 181.34 |
| 21. | Gschwender | 14 | 31m 29.41 | 112.44 | 181.07 |
| 22. | Franz | 14 | 31m 36.23 | 112.04 | 180.42 |
| 23. | Fischer | 14 | 31m 37.17 | 111.98 | 180.33 |
| 24. | Moineau | 14 | 31m 38.00 | 111.94 | 180.25 |
| 25. | Krummenacker | 14 | 32m 18.16 | 109.62 | 176.52 |
| 26. | Fahr | 14 | 32m 23.63 | 109.31 | 176.02 |
| 27. | Heiler | 13 | 31m 00.71 | 106.02 | 170.73 |

**NOT CLASSIFIED**

Iddon 12 laps, Fabbri 6, Burnett 5, Rubatto 5, Weibel 5, Monsch 3, Monti 1, Tardozzi 0, Mounier 0

## Championship Positions

**1.** Mertens 195, **2.** Merkel 192, **3.** Roche 172, **4.** Pirovano 165, **5.** Falappa 139, **6.** Andersson 121, **7.** Rymer 101, **8.** Baldwin 74, **9.** Monti 71,**10.** Vieira 51, **11.** Suhonen 46, **12.** Igoa 45, **13** McMurter 38, **14.** Dowson 37, **15.** Burnett 36, **16.** Polen 33, **17.** Tardozzi, Campbell 31, **19.** McElnea, Gschwender 28, **21.** Broccoli, Iwahashi, Hofmann 27, **24.** Rasmussen 26, **25.** Phillis 25, **26.** Douglas 20, **27.** Arata 18, **28.** Biliotti, Morrison, Rubatto 17, **31.** Goodfellow, Slight 16, **33.** Mounier 15, **34.** Iddon, Bolle 14, **36.** Hislop 13, **37.** Fogarty 12, **38.** Crine, Phillips, Crevier, Sohwa 11, **42.** Tsukamoto, Schmidt 10, **44.** Mercier, Delaby, Coq 7, **47.** Fabbri, Abe, Lucchinelli 6, **50.** McGladdery, Weibel 5, **52.** von Muralt, Ricci, de Vries, Mark 3, **56.** Moineau, Doohan, Kobayashi, Bouheben 2, **60.** Russell, Hamdi, Delcamp, Takeuchi, McGregor 1

# Round 9

# Pergusa, Sicily

## 24 September

*The Bimota truck was loaded with some expensive scrap metal after Falappa had crashed one of the team bikes in the Italian national race prior to the Superbike feature.* ***(Below)*** *Falappa needed every bit of encouragement from his pit crew to overcome the pain of his injured arm and make the World Superbike start.*

*(Left - Right) Merkel, Mertens and Pirovano celebrate in style. Mertens* ***(Below)*** *lost the points lead thanks to a misfire in the second race.*

Enna-Pergusa is located right in the middle of the island of Sicily, about seventy miles south west of Palermo. The circuit is not a memorable one being flat and consisting basically of four straights linked by four chicanes. Right in the centre of the facility is a stagnant lake.

Early in the season one of Fred Merkel's spannermen, Norris Farrow, said of Pergusa, "no one likes going there. It's a hell hole of a place, hot, humid and full of mosquitos."

After two bitterly fought races on the 3.6 mile track, Fred Merkel might have disagreed with Farrow's sentiments. After losing the points advantage with a lacklustre display in West Germany, the defending champion fought back into contention, scoring a pair of second places and in so doing, moved back into the lead of the 1989 World Superbike Championship.

***ENTRY/QUALIFYING***

Sicily is a long way to travel for even the well heeled European squads so it was no surprise several of the teams opted out of the trek, but the racing was a hard fought as ever in this second season of World Superbike.

It was a brave Giancarlo Falappa who grabbed the pole. The factory Bimota star had fallen the previous day in an Italian National Championship round and injured his shoulder, but as he proved in Austria when he rode with a knee injury, as long as he was capable of sitting on the bike, he would give it his best shot.

The front row pleased the partisan crowd with two Italian riders and four Italian machines completing the line up. The remainder of the row comprised Merkel on the Italian sponsored HRC Honda (complete with Italian made Pirelli tyres!), Falappa's teammate Baldwin, Ducati rider Monti and Belgian interloper Mertens on his rapid D'Holda Honda, another of the HRC supplied machines.

***RACE 1***

Only thirty two riders gridded for the first encounter and it was Pirovano who jetted into the lead. He was pursued by Monti, Mertens, Merkel and Falappa and the pace was frantic from the off with Andersson, Roche, Tardozzi and Baldwin completing the frontrunning bunch.

Two laps into the race the championship leading duo of Merkel and Mertens sped by Pirovano and forged a substantial lead. Merkel had his head down and was going for it but the crafty Mertens was happy to sit in his slipstream and choose his moment.

On the thirteenth lap, Mertens made his bid for the lead, pulling out of the American's draft down the back straight. Merkel tried desperately to find a way by in the remaining laps but his adversary held the advantage to the line and won by just under two seconds.

Merkel said, "that has to be one of the hardest races I've ever done. Stephane rode really hard, but fair. I was a bit quicker than him on the twisty sections but he was quicker at the back of the circuit. I'm gonna beat him in the second one though, just you wait and see."

Pirovano was a distant third, over thirty seconds off the pace but was six seconds clear of fourth placeman Monti. He inherited the place form his own teammate Roche who went out when the Ducati expired thanks to the failure of the one of the electrical pick ups.

Fifth place Anders Andersson was a further six seconds adrift while Jari Suhonen, on the Arwidson Yamaha just edged out Patrick Igoa and the Kawsasaki France ZXR750 for sixth place.

***RACE 2***

Heavy dark clouds hung over the circuit as the field blasted off towards the chicane for the start of the second leg. The head of the field was all red with the Ducati team of Monti and Roche leading the way.

Behind than came Baldwin, the sole representative of the Rimini based Bimota team as Falappa, in agony from his injured shoulder, opted to sit this one out.

*Patrick Igoa managed to get the French Kawasaki into the points again.*

*Monti was impressive in front of the Sicilian crowds on the factory Ducati.*

Merkel and Mertens were at it again, continuing their bitter on track feud but this time back in fifth and sixth places and chased by Scandinavian Yamaha pair Andersson, Suhonen and Frenchman Christophe Bouheben on the Kawasaki France ZXR750.

On the fifth lap, Baldwin went into the lead but it was shortlived and a lap later it was Mertens on a charge. He came under pressure from the two Ducatis and just when it looked likely they would pass him, the Belgian's bike faltered and he dropped to fifth with the vee four motor only firing on three of the cylinders.

Amidst this drama, Roche powered past his teammate and sped off into the distance, assuming total control over the race for a six second victory. Behind him, Monti and Merkel were embroiled in a scrap for second clinched by Merkel three laps from the end when he outbraked the less experienced rider going into the final chicane.

Merkel, having been held up by Monti for several laps, set off after Roche but once he realised there were too few laps to catch the flying Frenchman, backed off content with his second place, especially in the knowledge that Mertens had faded to seventh place.

Monti took a lone third place but Pirovano just pipped Suhonen and Igoa to the line for fourth place after an epic battle. Mertens took seventh with Bouheben completing a good day for the Kawasaki camp by finishing eighth, just ahead of Italian Bimota privateer, Marino Fabbri.

In terms of the championship, the result meant that Merkel regained his series lead, but only stood two points clear of Mertens and Pirovano ousted Roche (his first race dnf costing him dear in the point standings) for third by a single point. By now Falappa was out of contention, and with Bimota's decision not to contest any more races, out of the series, and only had seven points on sixth placeman Andersson.

They say it is tough at the top. With two rounds remaining in Australasia, it was going to get a lot tougher.

# Pergusa

*The race for the championship was never more obvious that at Pergusa between Merkel (1) and Mertens (4). Merkel edged ahead in the points in Sicily. Mertens hoped for better luck down under.*

NOLAN
1
Castrol

# Round 9

# Pergusa, Sicily

## 24 September

# Race Data

# Qualifiers

| POS | RIDER/NATIONALITY | ENTRANT/BIKE | TIME |
|---|---|---|---|
| 1. | Giancarlo Falappa (I) | Bimota Spa YB4EI | 1m 43.19 |
| 2. | Fred Merkel (USA) | RCM/Rumi Honda RC30 | 1m 44.18 |
| 3. | Mike Baldwin (USA) | Bimota YB4E1R | 1m 44.39 |
| 4. | Baldassarre Monti (I) | Squadra Corse Ducati 851S | 1m 44.45 |
| 5. | Stephane Mertens (B) | Team Mediares Honda RC30 | 1m 44.85 |
| 6. | Raymond Roche (F) | Squadra Corse Ducati 851S | 1m 44.90 |
| 7. | Anders Andersson (S) | Yamaha Sweden FZR750R | 1m 45.27 |
| 8. | Fabrizio Pirovano (I) | Belgarda Yamaha FVR750R | 1m 45.53 |
| 9. | Jari Suhonen (SF) | Arwidson Yamaha FVR750R | 1m 45.97 |
| 10. | Patrick Igoa (F) | Kawasaki France ZXR750 | 1m 46.02 |
| 11. | Davide Tardozzi (I) | Speed Shadows Bimota YB4E1 | 1m 46.38 |
| 12. | Arnad Harmati (H) | Honda VFR750 | 1m 46.88 |
| 13. | Marco Lucchinelli (I) | Squadra Corse Ducati 851S | 1m 47.03 |
| 14. | Christophe Bouheben (F) | Kawasaki France ZXR750 | 1m 47.56 |
| 15. | Rene Rassmussen (DK) | Pepsi Suzuki GSXR750R | 1m 47.57 |
| 16. | Oscar La Ferla (I) | Yamaha FVR750R | 1m 48.19 |
| 17. | Marco Dall'Aglio (I) | Yamaha FVR750 | 1m 48.29 |
| 18. | Marino Fabbri (I) | RB Moto Kawasaki ZXR750 | 1m 48.34 |
| 19. | Giorgio Cantalupo (I) | Ducati 851 | 1m 48.37 |
| 20. | Piergiorgio Bontempi (I) | RCM/Rumi Honda RC30 | 1m 48.43 |
| 21. | Aldeo Presciutti (I) | Yamaha | 1m 40.70 |
| 22. | Mauro Ricci (I) | Ducati 851 | 1m 48.73 |
| 23. | Corrado Manici (I) | Ducati | 1m 49.65 |
| 24. | Roberto Biagioli (I) | Yamaha FVR750R | 1m 50.48 |
| 25. | Silvano Ricchetti (I) | Honda RC30 | 1m 52.02 |
| 26. | Adriano Narducci (I) | Honda RC30 | 1m 54.27 |
| 27. | Alan Benedetti (I) | Yamaha | 1m 54.32 |
| 28. | Emanuele Battaglia (I) | Yamaha | 1m 55.05 |
| 29. | Antonio Mancuso (I) | Suzuki | 1m 57.90 |
| 30. | Sebastiano Zerbo (I) | Yamaha | 1m 58.06 |
| 31. | S. Pennisi (I) | Honda RC30 | 1m 58.11 |
| 32. | Alois Hager (A) | - | 1m 58.37 |
| 33. | Fabio Di Pietro (I) | Honda | 2m 00.25 |
| 34. | Mario Intelisano (I) | Ducati | 2m 02.91 |
| 35. | Biagio Di Dio (I) | Suzuki | 2m 04.17 |
| 36. | Francesco Garozzo (I) | Honda RC30 | 2m 04.30 |
| 37. | Giuseppe Coniglianoe (I) | Ducati | 2m 04.65 |
| 38. | Salvatore Calabrese (I) | Yamaha | 2m 06.28 |
| 39. | Stefano Scordato (I) | Suzuki | 2m 09.29 |
| 40. | Antonio Lo Cicero (I) | Suzuki | 2m 19.99 |

**Race Data:** 2 x 21 laps of 3.07 mile circuit
**Total Distance:** 128.94 miles
**Weather:** Sunny
**Fastest Lap:** Raymond Roche (race two) 1m 43.143s, 107.28 mph, 172.769 kph
**Existing Lap Record:** Establishes new class record

## Results Race 1

| POS | RIDER | LAPS | TIME | MPH | KPH |
|---|---|---|---|---|---|
| 1. | Mertens | 21 | 36m 46.90 | 105.30 | 169.57 |
| 2. | Merkel | 21 | 36m 47.10 | | |
| 3. | Pirovano | 21 | 37m 22.91 | | |
| 4. | Monti | 21 | 37m 30.73 | | |
| 5. | Andersson | 21 | 37m 36.71 | | |
| 6. | Suhonen | 21 | 37m 40.56 | | |
| 7. | Igoa | 21 | 37m 40.86 | | |
| 8. | Tardozzi | 21 | 37m 49.24 | | |
| 9. | Lucchinelli | 21 | 37m 55.64 | | |
| 10. | Bontempi | 21 | 38m 26.95 | | |
| 11. | Harmati | 21 | 38m 27.98 | | |
| 12. | Manici | 21 | 38m 33.63 | | |
| 13. | Presciutti | 20 | 37m 56.19 | | |
| 14. | Ricchetti | 20 | 38m 03.41 | | |
| 15. | Narducci | 20 | 38m 20.71 | | |
| 16. | Pennisi | 19 | 37m 08.30 | | |
| 17. | Zerbo | 19 | 37m 42.77 | | |
| 18. | Mancuso | 18 | 37m 23.18 | | |

**NOT CLASSIFIED**

Cantalupo 14 laps, Baldwin 9, Falappa 7, Fabbri 6, Bouheben 5, La Ferla 3, Roche 2, Hager 2, Ricci 0, Rasmussen 0, Dall'Aglio 0, Biagioli 0, Benedetti 0

***No further information supplied by race organisers***

## Results Race 2

| POS | RIDER | LAPS | TIME | MPH | KPH |
|---|---|---|---|---|---|
| 1. | Roche | 21 | 36m 54.61 | 104.93 | 168.98 |
| 2. | Merkel | 21 | 37m 00.46 | | |
| 3. | Monti | 21 | 37m 00.86 | | |
| 4. | Pirovano | 21 | 37m 27.02 | | |
| 5. | Suhonen | 21 | 37m 27.43 | | |
| 6. | Igoa | 21 | 37m 27.88 | | |
| 7. | Mertens | 21 | 37m 35.37 | | |
| 8. | Bouheben | 21 | 37m 51.90 | | |
| 9. | Fabbri | 21 | 37m 52.27 | | |
| 10. | Dall'Aglio | 21 | 38m 06.06 | | |
| 11. | Harmati | 21 | 38m 11.49 | | |
| 12. | Bontempi | 21 | 38m 15.71 | | |
| 13. | Presciutti | 21 | 38m 23.15 | | |
| 14. | Manici | 21 | 38m 23.34 | | |
| 15. | La Ferla | 21 | 38m 33.56 | | |
| 16. | Ricci | 21 | 38m 34.51 | | |
| 17. | Cantalupo | 21 | 38m 34.82 | | |
| 18. | Ricchetti | 20 | 37m 58.61 | | |
| 19. | Pennisi | 19 | 37m 35.62 | | |
| 20. | Zerbo | 19 | 37m 59.31 | | |
| 21. | Mancuso | 19 | 38m 32.52 | | |

**NOT CLASSIFIED**

Andersson 14 laps, Tardozzi 8, Baldwin 7, Battaglia 3, Rasmussen 1, Hager 1, Benedetti 0, Narducci 0, Falappa 0, Lucchinelli 0, Biagioli 0

## Championship Positions

**1.** Merkel 226, **2.** Mertens 224, **3.** Pirovano 193, **4.** Roche 192, **5.** Falappa 139, **6.** Andersson 132, **7.** Rymer 101, **8.** Monti 99, **9.** Baldwin 74,**10** Suhonen 67, **11.** Igoa 64, **12.** Vieira 51, **13.** Tardozzi 39, **14.** McMurter 38, **15.** Dowson 37, **16.** Burnett 36, **17.** Polen 33, **18.** Campbell 31, **19.** McElnea, Gschwender 28, **21.** Broccoli, Iwahashi, Hofmann 27, **24.** Rasmussen 26, **25.** Phillis 25, **26.** Douglas 20, **27.** Arata 18, **28.** Biliotti, Morrison, Rubatto 17, **31.** Goodfellow, Slight 16, **33.** Mounier 15, **34.** Iddon, Bolle 14, **36.** Hislop, Fabbri, Lucchinelli 13, **39.** Fogarty 12, **40.** Crine, Phillips, Crevier, Sohwa 11, **44.** Tsukamoto, Schmidt, Bontempi, Harmati, Bouheben 10, **49.** Manici 8, **50.** Mercier, Delaby, Coq 7, **53.** Presciutti, Dall'Aglio 6, **55.** McGladdery, Weibel 5, **57.** von Muralt, Ricci, de Vries, Mark 3, **61.** Moineau, Doohan, Kobayashi, Richetti 2, **65.** Rusell, Hamdi, Delcamp, Takeuchi, McGregor, Narducci, La Ferla 1.

# Round 10

## Oran Park, Australia

## 12 November

*Marlboro Yamaha Dealer Team Australia in strength on home soil. Up and coming Rene Bongers lines up with Dowson (centre) and Peter Goddard.*

*A view of the Oran Park pit 'facility'.* ***(Bottom Left)*** *Pirovano takes a time out before practice begins at Oran.*

*Dowson (23), Merkel (1), Campbell (9), Rymer (partially obscured by Campbell), and Mertens (4) get away in the difficult conditions at Oran.*

The Australian Marlboro Yamaha dealer team of Peter Goddard and Michael Dowson dominated the penultimate round of the World Superbike Championship on their home soil at Oran Park, Sydney but the leading protagonists in the title race, Stephane Mertens and defending champion Fred Merkel continued their almost season long fight.

Neither rider shone as far as overall results were concerned but Merten's 8-4 finishes were enough to return the points lead to the Belgian corner as Merkel faltered with 11-5 finishes.

***ENTRY/QUALIFYING***

Only 36 riders entered for the penultimate round of the championship at Oran Park. Missing were several of the top European teams. No one expected teams like Kawasaki France and Suzuki France to make the effort after they had only raced in selected European races as their budgets allowed.

However, it was a great disappointment that the Bimota pairing of Giancarlo Falappa and Mike Baldwin were absent and Ducati only sent one bike for Raymond Roche a move which denied their promising Baldassarre Monti valuable overseas experience.

The entry still had a quality look to it, even if the quantity was lacking. In addition to the leading Superbike contestants, the home entry featured, Sugo winner Michael Dowson plus his Marlboro Yamaha Dealer team partner Peter Goddard.

Goddard, a twenty five year old from Narellan, NSW, came into the race full of confidence after winning the Tsubaka All Japan Formula One race only days before. It marked the first international win for the former four time Australian dirt track champion.

With 29 year old Dowson alongside him on the same trick lightweight machine he had raced to victory in Sugo, Marlboro Yamaha were hoping for a repeat of their 1988 World Superbike Oran Park result when Michael Doohan (now in the Rothmans Honda GP500 squad) teamed with Dowson for a memorable 1-2 in each of the races. The team were also blooding fast rising Rene Bongers, a promising 250 rider from Victoria.

Other leading locals included; Team Kawasaki Australia representatives, gritty Rob Phillis (with a specially flown in factory bike from Japan) and young Aaron Slight, Malcolm Campbell was on the Honda

Performance Tyres

*Mertens outscored Merkel at Oran to leapfrog back into the points lead.*

TOTAL
HONDA
mediares
m team
BEL RAY
ÖHLINS
AFAM
MJK
LAZER

Australia RC30 and 250 Australian champion Darryl Beattie was being given an outing on the 008 Telecom Readyline RC30 normally raced by Sean Gallagher.

Beattie is regarded as potentially the next superstar to come out of Australia. Just 19, he had already wrapped up the Australian national title riding an NSR250 Honda and going into Oran Park was celebrating the offer of a works Honda contract in the Formula One class which would see him racing in Japan on a regular basis during 1990.

Also in the line up were Kawasaki Japan's official factory rider Takahiro Sohwa and multi-Canadian Champion Steve Crevier on a Yamaha.

Qualifying was dogged by unexpected poor weather conditions which probably contributed to the closeness of the practice times.

Dowson predictably grabbed pole with a 1m 13.226s but fellow Aussie Malcolm Campbell was also hot on the pace only three tenths of a second behind.

Stephane Mertens was the quickest of the Superbike regulars in third place with a 1m 14.054s and completing the front row were Roche, Goddard, who was bruised and battered after he had crashed on cold tyres during Friday practice, and Merkel. Next came the two Aussie Kawasakis of Phillis and Slight respectively.

***RACE 1***

The first race was delayed by fifteen minutes to see if the falling rain would stop or worsen. In the end officials decided to proceed with the meeting and race one was started with a steady drizzle wetting the track.

Everyone opted for wet tyres except Goddard who fitted slicks to his OWO1. As the field splashed away from the green light, it looked like Goddard had made a costly error.

In front of a poor crowd of just 5,000 people, Goddard's teammate, Dowson, led the charge with British rider Terry Rymer (back in the Superbike fray after missing the past two rounds due to British Championship commitments with his Loctite Yamaha team) Phillis and Pirovano in the wake of the leading red and white Marlboro Yamaha.

For five laps Goddard slithered around on his slicks but suddenly the rain stopped and the sun appeared to dry the track surface. Shortly after Rymer lost it when he strayed from the wet line onto the dry portion of the track.

At the mid point of the race, Phillis forced his ZXR750 by Dowson for the lead and looked to be in control for a victory aboard the green machine.

Behind Phillis came Dowson who was under increasing pressure from Pirovano and Slight with Goddard starting make his play for the leaderboard placings.

As the track surface got drier, so Goddard increased his speed and was soon at the head of the field and pulling clear. No one could live with him and the more pressure the pursuers applied the worse their tyre situation became as the deep cut treaded tyres overheated.

At the flag, Goddard won by over four seconds from Phillis, Pirovano and Roche. Dowson faded to seventh. Mertens got eighth but Merkel was in trouble back down in eleventh place, although his team protested an alleged illegal swing arm fitted to the tenth placed Suzuki of Scott Doohan (Michael's brother).

***RACE 2***

Just prior to the start of the second leg the feared drizzle began again but thankfully was not enough to dampen the track and everyone started the race on slicks.

Dowson won the holeshot and immediately pulled out a clear advantage. Campbell, Pirovano, Goddard, Rymer and Mertens led the pursuers but Dowson's pace was so hot, no one was going to get onto the back wheel of the Marlboro Yamaha this time.

The interest in the race for the small crowd centred on two separate clashes. Pirovano was battling with Campbell for third while Mertens and Merkel assumed their now regular task of shadowing one another back in a fight for seventh place.

Roche, overcoming one of his customary

*World Superbike or not, Aussies Rob Phillis (left) ar*

*Dowson had some scores of their own to settle. Both had special factory bikes direct from Japan for the meeting.*

bad starts was up to second but nothing he could do, even on the rapid vee twin Ducati, was making any impression on the leader, Dowson.

Goddard was one of the riders Roche passed in his charge and as the Aussie tried to keep with him, he crashed, for the second time in the weekend. Subdued, but otherwise unhurt, Goddard was man enough to admit it was simply a case of pilot error, although he did add his Yamaha was way down on horsepower compared with Roche's fire engine.

Three quarters into the race Campbell ran into the rear end of Pirovano's Yamaha and the resulting crash put both of them out of the running for third place. That put Phillis in a handy lone third place but the collision spelt disaster for Terry Rymer who fell on the spilt oil from the crash. He was taken to hospital for treatment to an injured hand.

The race long Mertens/Merkel duel was resolved in Merten's favour and he beat the American to the flag by less than half a second to establish a narrow three point lead going into the final round at Manfield, New Zealand the following weekend.

# Round 10

# Oran Park, Australia

## 12 November

## Race Data

# Qualifiers

| POS | RIDER/NATIONALITY | ENTRANT/BIKE | TIME |
|---|---|---|---|
| 1. | Michael Dowson (AUS) | Yamaha Australia FZR750R | 1m 13.226 |
| 2. | Malcolm Campbell (AUS) | Citizen Honda RC30 | 1m 13.596 |
| 3. | Stephane Mertens (B) | Team Mediares Honda RC30 | 1m 14.054 |
| 4. | Raymond Roche (F) | Squadra Corse Ducati 851S | 1m 14.060 |
| 5. | Peter Goddard (AUS) | Yamaha FZR750R | 1m 14.173 |
| 6. | Fred Merkel (USA) | RCM/Rumi Honda RC30 | 1m 14.325 |
| 7. | Rob Phillis (AUS) | Team Kawasaki Aus/ZXR750 | 1m 14.407 |
| 8. | Aaron Slight (NZ) | Team Kawasaki Aus/ZXR750 | 1m 14.527 |
| 9. | Scott Doohan (AUS) | Suzuki GSXR750R | 1m 14.746 |
| 10. | Rene Bongers (AUS) | Yamaha FZR750 | 1m 14.760 |
| 11. | Terry Rymer (GB) | Loctite Yamaha FZR750R | 1m 14.769 |
| 12. | Fabrizio Pirovano (I) | Belgarda Yamaha FZR750R | 1m 14.809 |
| 13. | Anders Andersson (S) | Yamaha Sweden FZR750R | 1m 15.108 |
| 14. | Robert Scolyer (AUS) | Honda RC30 | 1m 15.183 |
| 15. | Daryl Beattie (AUS) | Honda RC30 | 1m 15.318 |
| 16. | Jari Suhonen (S) | Arwidson Yamaha FZR750R | 1m 15.329 |
| 17. | Graeme Morris (AUS) | Ducati | 1m 15.415 |
| 18. | Takahiro Sohwa (J) | KRT Kawasaki ZXR750 | 1m 15.558 |
| 19. | Robert Holden (NZ) | Bob Brown Ducati | 1m 15.848 |
| 20. | Matthew Blair (AUS) | Suzuki GSXR750R | 1m 15.849 |
| 21. | Roy Leslie (AUS) | Yamaha FZR750R | 1m 15.985 |
| 22. | Gary Goodfellow (CDN) | Don Knit/Yoshimura Suzuki | 1m 16.129 |
| 23. | Steve Crevier (CDN) | Yamaha FZR750R | 1m 16.4223 |
| 24. | Andy Roberts (AUS) | Yamaha FZR750R | 1m 16.698 |
| 25. | Steve Martin (AUS) | Suzuki GSXR750R | 1m 16.828 |
| 26. | Andy McGladdery (GB) | Francis Neill Honda RC30 | 1m 16.883 |
| 27. | Peter Guest (AUS) | Suzuki GSXR750R | 1m 17.015 |
| 28. | David Ross (AUS) | Suzuki GSXR750R | 1m 17.089 |
| 29. | Roger Heyes (AUS) | Yamaha FZR750R | 1m 17.100 |
| 30. | John Richards (AUS) | Yamaha FZR750R | 1m 17.200 |
| 31. | Russell Davis (AUS) | Kawasaki ZXR750 | 1m 17.935 |
| 32. | Peter Salmon (AUS) | Honda RC30 | 1m 18.045 |
| 33. | Chris Oldfield (AUS) | Bimota YB4 | 1m 18.745 |
| 34. | Scott Mitchell (AUS) | Yamaha FZR750R | 1m 19.120 |
| 35. | Daniel Vanolini (AUS) | Honda RC30 | 1m 19.587 |
| 36. | Chris K. Oldfield (AUS) | Ducati | 1m 21.314 |

**Race Data:** 2 x 39 laps of 1.62 mile circuit
**Total Distance:** 126.36 miles
**Weather:** Rain/Fine
**Fastest Lap:** Michael Dowson (race two) 1m 13.215s, 80.00 mph, 128.826 kph
**Existing Lap Record:** Fred Merkel 1m 12.690, 80.57 mph, 129.75 kph (September 1988)

# Results Race 1

| POS | RIDER | LAPS | TIME | MPH | KPH |
|---|---|---|---|---|---|
| 1. | Goddard | 39 | 52m 55.552 | | |
| 2. | Phillis | 39 | 53m 51.808 | | |
| 3. | Pirovano | 39 | 54m 05.537 | | |
| 4. | Roche | 39 | 54m 12.297 | | |
| 5. | Slight | 38 | 53m 01.966 | | |
| 6. | Andersson | 38 | 53m 13.872 | | |
| 7. | Dowson | 38 | 53m 14.337 | | |
| 8. | Mertens | 38 | 53m 14.574 | | |
| 9. | Bongers | 38 | 53m 26.025 | | |
| 10. | Doohan | 38 | 53m 43.175 | | |
| 11. | Merkel | 38 | 53m 43.659 | | |
| 12. | Suhonen | 38 | 53m 49.462 | | |
| 13. | Richards | 38 | 53m 00.409 | | |
| 14. | Sohwa | 38 | 54m 00.877 | | |
| 15. | Martin | 38 | 54m 02.512 | | |
| 16. | Blair | 37 | 53m 09.743 | | |
| 17. | Salmon | 37 | 53m 11.191 | | |
| 18. | Campbell | 37 | 53m 29.736 | | |
| 19. | Roberts | 37 | 53m 47.866 | | |
| 20. | Scolyer | 37 | 54m 01.912 | | |
| 21. | Mitchell | 37 | 54m 19.877 | | |
| 22. | Oldfield CK | 36 | 53m 34.631 | | |
| 23. | Heyes | 36 | 53m 36.832 | | |
| 24. | Davis | 34 | 54m 00.161 | | |
| 25. | Ross | 33 | 53m 12.112 | | |
| 26. | Vanolini | 33 | 53m 15.215 | | |
| 27. | Goodfellow | 32 | 53m 41.548 | | |

**NOT CLASSIFIED**
McGladdery 32 laps, Morris 19, Beattie 15, Leslie 13, Oldfield 12, Guest 11, Holden 8, Rymer 6,

# Results Race 2

| POS | RIDER | LAPS | TIME | MPH | KPH |
|---|---|---|---|---|---|
| 1. | Dowson | 39 | 48m 5.321 | | |
| 2. | Roche | 39 | 48m 8.575 | | |
| 3. | Phillis | 39 | 48m 28.606 | | |
| 4. | Mertens | 39 | 48m 36.153 | | |
| 5. | Merkel | 39 | 48m 36.415 | | |
| 6. | Slight | 39 | 48m 38.375 | | |
| 7. | Sohwa | 39 | 48m 48.419 | | |
| 8. | Bongers | 39 | 49m 7.749 | | |
| 9. | Blair | 39 | 49m 10.399 | | |
| 10. | Morris | 39 | 49m 11.974 | | |
| 11. | Scolyer | 39 | 49m 12.244 | | |
| 12. | Crevier | 39 | 49m 14.307 | | |
| 13. | Roberts | 38 | 48m 33.953 | | |
| 14. | McGladdery | 38 | 48m 35.922 | | |
| 15. | Richards | 38 | 48m 42.789 | | |
| 16. | Heyes | 38 | 48m 50.028 | | |
| 17. | Beattie | 38 | 49m 43.080 | | |
| 18. | Mitchell | 37 | 48m 29.744 | | |
| 19. | Oldfield CK | 37 | 54m 57.829 | | |
| 20. | Davis | 35 | 55m 19.581 | | |
| 21. | Ross | 34 | 54m 30.933 | | |
| 22. | Vanolini | 34 | 54m 38.563 | | |
| 23. | Goodfellow | 33 | 54m 58.696 | | |

**NOT CLASSIFIED**
McGladdery 32 laps, Morris 19, Beattie 15,Leslie 13,Oldfield 12, Guest 11, Holden 8, Rymer 6

*No further information supplied by race organisers*

# Championship Positions

**1.** Mertens 245, **2.** Merkel 242, **3.** Roche 222, **4.** Pirovano 208, **5.** Andersson 142, **6.** Falappa 139, **7.** Rymer 101, **8.** Monti 99, **9.** Baldwin 74, **10.** Suhonen 73, **11.** Dowson 66, **12.** Igoa 64, **13.** Phillis 57, **14.** Vieira 51, **15.** Tardozzi 39, **16.** McMurter 38, **17.** Burnett 36, **18.** Polen 33, **19.** Campbell 31, **20.** McElnea, Gschwender 28, **22.** Broccoli, Iwahashi, Hofmann 27, **25.** Rasmussen 26, **26.** Sohwa 22, **27.** Douglas, Goddard 20, **29.** Arata 18, **30.** Biliotti, Morrison, Rubatto 17, **33.** Goodfellow, Slight 16, **35.** Mounier, Bongers, Crevier 15, **38.** Iddon, Bolle 14, **40.** Hislop, Fabbri, Lucchinelli 13, **43.** Fogarty 12, **44.** Crine, Phillips 11, **46.** Tsukamoto, Schmidt, Bontempi, Harmati, Bouheben 10, **51.** Nanici, Doohan 8, **53.** Mercier, Delaby, Coq, Blair, McGladdery 7, **58.** Presciutti, Dall'Aglio, Morris 6, **61.** Weibel, Scolyer 5, **63.** Richards 4, **64.** von Muralt, Ricci, de Vries, Mark, Roberts 3, **69.** Moineau, Kobayashi, Richetti 2, **72.** Russell, Hamdi, Delcamp, Takeuchi, McGregor, Narducci, La Ferla, Martin 1.

# Round 11

# Manfeild, New Zealand

## 19 November

*Mertens discusses the conditions with the Michelin representative before contemplating his final tyre choice.* ***(Below)*** *Pit conditions were primitive at Manfeild as Malcolm Campbell's team discover. Yes, that is a container the team are 'housed' in!*

*Team Kawasaki were also undecided about tyre choice.* ***(Top right)*** *Japanese Kawasaki teamster Sohwa failed to impress in Australasia.* ***(Below)*** *Merkel's mechanic, Chris Johnson gives the message that Mertens has crashed out.*

The championship, as everyone had expected, went to the wire and it was Flyin' Fred Merkel who came out on top for the second successive season of World Superbike.

The Californian rode his Rumi backed HRC Honda superbly, mastering the difficult conditions with a faultless pair of third place rides to clinch the title and give Pirelli tyres their first ever World Championship in motorcycle road racing.

A jubilant Merkel declared, "This is the happiest day of my life. The thrill is much better than last year. It's real tough winning two years running."

Merkel went into the Manfeild race trailing Belgian Stephane Mertens by three points but he had seen his chief rival crash heavily in the first race and knew consistent results in both legs would assure him the title.

***ENTRY/QUALIFYING***

Merkel won the first part of the psychological battle when he posted pole time of 1m 08.33s, just one tenth of a second faster than Aussie Honda man Malcolm Campbell and local boy Aaron Slight.

Merkel had a changed attitude after Oran Park where he failed to get on the pace. After qualifying Merkel said, "For me, pole position is psychologically very important. I was in this position last year so I know that I have to go for it. I'm gonna ride every corner as if it was a new experience. I'm not interested in riding for points, I'm going out to win both races. I've got one really good tyre which I'm gonna use in the first race and one slightly harder one for the second. They'll both last the distance okay."

Slight was on his regular Team Kawasaki Australia mount but teammate Rob Phillis had the same factory bike from Japan that he had raced at Oran Park. He had qualified on the front row alongside fellow countryman

*Peter Goddard didn't enjoy the same degree of good fortune in New Zealand as he had done in Australia.*

Dowson and Mertens.

The points leader was apprehensive knowing he only held a slender three point cushion. His tuner, Jean d'Hollander admitted, "For sure it will be difficult. I am very nervous. We have worked hard all season and now it all depends on tomorrow."

All the front row clocked 1m 8's, as did first of the second row men, Englishman Terry Rymer, the battered Loctite Yamaha rider still recuperating from his Oran Park endo.

Next up came Canadian domiciled Kiwi Gary Goodfellow with his Don Knit Sugano mount Raymond Roche on the sole factory Ducati, another Canadian, multi champion Steve Crevier on a Yamaha, and Swede Anders Andersson with his rapid OWO1.

***RACE 1***

Faced with a damp track most riders opted for wets as they lined up for the first race. Rymer holeshot the start from the second row and led the pack around for the opening lap.

Soaking up his spray were Dowson, Phillis, Slight, Suhonen and Merkel then as Rymer forged clear he briefly took Dowson with him until the experimental Dunlops on the British Yamaha reached working temperature and Rymer was able to make a clean break.

Mertens had not been impressive at the start but had fought his way through the pack and was up to sixth and looking good when disaster struck the championship aspirant.

On the twenty second lap, approaching Higgins Bend, the front wheel suddenly locked up and he was thrown off the bike. He was lucky to escape with a shaking but the bike cleared the gravel track and a row of small trees before smashing into a low and bursting into flames.

"For the two laps before, I felt a vibration when I braked," said the dejected Belgian, "and I nearly lost the bike at the hairpin. I was braking for the bend just before the back straight (Higgins) and the next thing I knew, I was on the ground. We think the front disc

*The remains of Mertens Honda front end after the crash which ultimatley lost him the championship.* ***(Below)*** *Merkel in his usual stylish, yet aggressive pose.*

SHARP COPIERS
7
10
agv

broke up and locked the wheel. I am very lucky not to have hurt myself.''

Although he knew Mertens was out, Merkel did not ease his pace and swept past Phillis, Suhonen and Dowson to take third place, four seconds down on the local boy, Slight.

The Kiwi was delighting the meagre 5,000 who had braved the unseasonal weather, but even he was powerless to stop Rymer's winning ways and wound up second, over 14 seconds adrift.

As Rymer passed the chequered flag, he punched the air in delight before acknowledging the applause of the crowd. On the rostrum he beamed, ''I really needed that. Especially after last week at Oran Park. I planned to pull away at the front but as the race went on it was hard to keep the pace up. It was good that it was wet out there. It meant I didn't have to push the bike too hard. I just wish my dad could have been here to see this.''

***RACE 2***

The track was dry as the field lined up for the second leg, but before the lights turned green a slight drizzle had begun to fall. Mertens was back in the line up on his spare machine, but barring a disaster for Merkel had little chance of the title now.

Everyone stayed on slicks and it was Dowson leading the charge from Slight, a determined looking Mertens, Rymer, Campbell, and Phillis.

Fifteen laps into the 33 lapper, Mertens took command of the race from Dowson and quickly established complete control. Even Dowson looked to have settled for a placing when he inexplicably slid off.

''Rymer had gone and I knew I wasn't catching him, so I was just cruising around. I guess I wasn't looking where I was going and I touched the white line,'' said a rueful Dowson. ''I was pitched off which is a bit stupid but it was my own fault.''

Mertens was unassailable up front but behind him Campbell, Merkel, Rymer, Goddard and Phillis were going at it hard. Campbell was determined to make amends after a first race crash and was giving it everything.

His pace took him, with Merkel in tow, clear of the group leaving Rymer, Goddard and Phillis to battle out fourth place. Goddard, after retiring from the first leg with mechanical problems, crashed with four laps to go just after he moved into fourth spot. Rymer was lucky to avoid the errant machine as it spun across the track in front of him but held off Phillis at the flag.

As Mertens ran out an easy winner, by nearly thirty eight seconds, Campbell beat Merkel home for his best result of the year. Merkel meantime was content with third place and the title.

Merkel and Mertens fought each other tooth and nail all season long but on the day it mattered, good ol' Flyin' Fred had ridden magnificently for his and Honda's second Superbike title and Pirelli their first World title.

As he savoured the winner's champagne Merkel said that if he didn't get a top flight GP ride in 1990 he would be back to defend his title for a third year. Two World Superbike crowns, AMA Superbike champion three times. Can anyone stop the man they have dubbed, 'Mr. Superbike'?

*Everyone marvelled at the performance of Briton, Terry Rymer in the damp first race at Manfeild. He simply motored away from the pack.*

ROB
SHOEI
TEAM
Kawasaki
AUSTRALIA
Kawasaki

*Riding a trick factory ZXR750, Rob Phillis top scored at Oran Park and featured strongly at Manfeild. A full season of World Superbike in 1990 will see him become a major threat for the title.*

# Qualifiers

# Round 11

# Manfeild, New Zealand

## 19 November

# Race Data

| POS | RIDER/NATIONALITY | ENTRANT/BIKE | TIME |
|---|---|---|---|
| 1. | Fred Merkel (USA) | RCM/Rumi Honda RC30 | 1m 08.33 |
| 2. | Malcolm Campbell (AUS) | Citizen Honda RC30 | 1m 08.46 |
| 3. | Aaron Slight (NZ) | Team Kawasaki Aus/ZXR750 | 1m 08.57 |
| 4. | Michael Dowson (AUS) | Yamaha Australia FZR750R | 1m 08.65 |
| 5. | Stephane Mertens (B) | Total/Bel Ray Honda | 1m 08.75 |
| 6. | Rob Phillis (AUS) | Team Kawasaki Aus/ZXR750 | 1m 08.75 |
| 7. | Terry Rymer (GB) | Loctite Yamaha FZR750R | 1m 08.93 |
| 8. | Gary Goodfellow (CDN) | Don Knit Yoshimura Suzuki | 1m 09.08 |
| 9. | Raymond Roche (F) | Squadra Corse Ducati | 1m 09.18 |
| 10. | Steve Crevier (CDN) | Yamaha FZR750R | 1m 09.38 |
| 11. | Anders Andersson (S) | Yamaha Sweden FZR750R | 1m 09.55 |
| 12. | Andrew Stroud (NZ) | Yamaha FZR750R | 1m 09.64 |
| 13. | Takahiro Sohwa (J) | KRT Kawasaki ZXR750 | 1m 09.75 |
| 14. | Peter Goddard (AUS) | Yamaha FZR750R | 1m 09.75 |
| 15. | Jari Suhonen (SF) | Arwidson Yamaha FZR750 | 1m 09.93 |
| 16. | Simon Crafar (NZ) | Yamaha FZR750R | 1m 10.94 |
| 17. | Glenn Williams (NZ) | Ducati 851EVD | 1m 11.46 |
| 18. | Andy McGladdery (GB) | Francis Neill Honda RC30 | 1m 11.74 |
| 19. | Mike King (NZ) | Ducati 851 | 1m 12.31 |
| 20. | Dale Robinson (GB) | Honda RC30 750 | 1m 12.43 |
| 21. | Eddie Kattenberg (NZ) | Suzuki GSXR750 | 1m 12.54 |
| 22. | Brian Bernard (NZ) | Suzuki GSXR750R | 1m 12.61 |
| 23. | Dale Warren (NZ) | Honda VFR750R | 1m 12.95 |
| 24. | Brian Billet (NZ) | Honda RC30 750 | 1m 13.39 |
| 25. | Russell Josiah (NZ) | Kawasaki ZXR750 | 1m 14.19 |
| 26. | Grant Ramage (NZ) | Suzuki FZR750R | 1m 14.41 |
| 27. | Tony McMurdo (NZ) | Kawasaki GPX750 | 1m 15.16 |
| 28. | Steve Scullion (NZ) | Ducati 888 | 1m 15.66 |
| 29. | Scott Williams (NZ) | Yamaha FZR750R | 1m 16.07 |
| 30. | Alistair Stronach (NZ) | Suzuki GZXR750 | 1m 18.97 |

**Race Data:** 2 x 33 laps of 1.94 mile circuit
**Total Distance:** 128.04
**Weather:** Wet/dry
**Fastest Lap:** Aaron Slight (Race 2) 1m 08.17s
**Existing Lap Record:** Stephane Mertens 1m 11.26s (October 1988)

# Results Race 1

| POS | RIDER | LAPS | TIME | MPH | KPH |
|---|---|---|---|---|---|
| 1. | Rymer | 33 | 42m 38.08 | | |
| 2. | Slight | 33 | 42m 52.21 | | |
| 3. | Merkel | 33 | 42m 56.59 | | |
| 4. | Dowson | 33 | 42m 56.65 | | |
| 5. | Phillis | 33 | 42m 57.12 | | |
| 6. | Suhonen | 33 | 43m 05.51 | | |
| 7. | Andersson | 33 | 43m 14.34 | | |
| 8. | Williams | 32 | 42m 39.51 | | |
| 9. | Sohwa | 32 | 42m 42.22 | | |
| 10. | Crafar | 32 | 43m 03.33 | | |
| 11. | Kattenberg | 32 | 43m 30.59 | | |
| 12. | Josiah | 32 | 43m 32.96 | | |
| 13. | McGladdery | 32 | 43m 36.26 | | |
| 14. | Bernard | 31 | 44m 02.60 | | |
| 15. | Stroud | 30 | 43m 33.19 | | |

**NOT CLASSIFIED**
Campbell 29 laps, Ramage 28, Warren 28,Mertens 21, Goddard 21, Billet 13,Williams 7, Roche 5,King 5, Robinson 3, Goodfellow 0, Scullion 0, McMurdo 0,Stronach 0

*No further information supplied by race organisers*

# Results Race 2

| POS | RIDER | LAPS | TIME | MPH | KPH |
|---|---|---|---|---|---|
| 1. | Mertens | 33 | 38m 52.26 | | |
| 2. | Campbell | 33 | 39m 14.31 | | |
| 3. | Merkel | 33 | 39m 17.71 | | |
| 4. | Rymer | 33 | 39m 21.28 | | |
| 5. | Phillis | 33 | 39m 26.90 | | |
| 6. | Stroud | 33 | 39m 32.33 | | |
| 7. | Suhonen | 32 | 38m 56.17 | | |
| 8. | Andersson | 32 | 38m 57.67 | | |
| 9. | Sohwa | 32 | 39m 02.54 | | |
| 10. | Crafar | 32 | 39m 07.96 | | |
| 11. | McGladdery | 31 | 39m 04.82 | | |
| 12. | Ramage | 31 | 39m 07.33 | | |
| 13. | Kattenberg | 31 | 39m 26.27 | | |
| 14. | Josiah | 31 | 39m 41.27 | | |
| 15. | Williams | 30 | 39m 03.74 | | |
| 16. | Crevier | 29 | 38m 57.80 | | |
| 17. | King | 29 | 40m 00.60 | | |

**NOT CLASSIFIED**
Goddard 29 laps, Dowson 25, Warren 15,Slight 12, Robinson 11, Billet 9, Roche 7, Bernard 4, Williams G 3, Goodfellow 0, Scullion 0, McMurdo 0, Stronach 0

# Championship Positions

**1.** Merkel 272, **2.** Mertens 265, **3.** Roche 222, **4.** Pirovano 208, Andersson 159, **6.** Falappa 139, **7.** Rymer 134, **8.** Monti 99, **9.** Suhonen 90, **10.** Dowson, Phillis 79, **12.** Baldwin 74, **13.** Igoa 64, **14.** Slight 54, **15.** Vieira 51, **16.** Campbell 48, **17.** Tardozzi 39, **18.** McMurter 38, **19.** Burnett, Sohwa 36, **21.** Polen 33, **22.** McElnea, Gschwender 28, **24.** Iwahashi, Hofmann 27, **26.** Broccoli, Rasmussen 26, **28.** Goddard, Crevier, Douglas 20, **31.** Arata 18, **32.** Rubatto, Biliotti, Morrison 17, **35.** Goodfellow 16, **36.** Mounier Bongers 15, **38.** Bolle, Iddon, McGladdery 14, **41.** Hislop, Lucchinelli, Fabbri 13, **44.** Fogarty, Crafar 12, **46.** Crine, Phillips, Stroud 11, **49.** Bouheben, Bontempi, Harmati, Schmidt, Tsukamoto 10, **54.** Williams, Doohan 8, **56.** Blair, Coq, Mercier, Kattenberg, Delaby 7, **61.** Morris, Dall'Aglio, Abe, Manici, Presciutti 6, **66.** Scolyer, Josiah, Weibel 5, **69.** Ramage, Chandler, Richards 4, **72.** Roberts, Mark, de Vries, Ricci, von Muralt 3, **77.** Richetti, Kobayashi, Moineau 2, **80.** Williams, Bernard, Martin, Takeuchi, Hamdi, Scott, La Ferla, Narducci, McGregor, Russell 1

# Statistics 1989

# 1989 Superbike World Championship

| | | | | GB | H | CDN | USA | A | F | J | D | I | AUS | NZ | PTS |
|---|---|---|---|---|---|---|---|---|---|---|---|---|---|---|---|
| **1** | Fred Merkel | USA | Honda | 13/10 | 20/20 | 20/15 | 13/15 | 5/15 | 8/13 | 0/4 | 8/13 | 17/17 | 5/11 | 15/15 | **272** |
| **2** | Stephane Mertens | B | Honda | 0/0 | 11/10 | 13/11 | 15/17 | 15/20 | 20/15 | 8/8 | 17/15 | 20/9 | 8/13 | 0/20 | **265** |
| **3** | Raymond Roche | F | Ducati | 0/0 | 17/0 | 17/0 | 20/20 | 17/6 | 15/17 | 3/0 | 20/20 | 0/20 | 13/17 | 0/0 | **222** |
| **4** | Fabrizio Pirovano | I | Yamaha | 20/0 | 15/17 | 10/13 | 17/13 | 11/17 | 0/0 | 9/0 | 13/10 | 15/13 | 15/0 | 0/0 | **208** |
| **5** | Anders Andersson | S | Yamaha | 10/11 | 10/9 | 7/5 | 8/9 | 8/7 | 6/9 | 0/0 | 11/11 | 11/0 | 10/0 | 9/8 | **159** |
| **6** | Giancarlo Falappa | I | Bimota | 0/20 | 7/13 | 0/20 | 0/0 | 3/0 | 11/20 | 13/15 | 0/17 | 0/0 | 0/0 | 0/0 | **139** |
| **7** | Terry Rymer | GB | Yamaha | 15/17 | 9/7 | 0/8 | 11/11 | 7/0 | 13/0 | 0/3 | 0/0 | 0/0 | 0/0 | 20/13 | **134** |
| **8** | Baldassarre Monti | I | Ducati | 0/0 | 0/11 | 0/9 | 10/10 | 10/2 | 9/10 | 0/0 | 0/0 | 13/15 | 0/0 | 0/0 | **99** |
| **9** | Jari Suhonen | F | Yamaha | 0/0 | 2/5 | 9/7 | 0/0 | 6/9 | 0/0 | 0/0 | 6/2 | 10/11 | 4/0 | 10/9 | **90** |
| **10** | Michael Dowson | AUS | Yamaha | 0/0 | 0/0 | 0/0 | 0/0 | 0/0 | 0/0 | 17/20 | 0/0 | 0/0 | 9/20 | 13/0 | **79** |
| **11** | Rob Phillis | AUS | Kawasaki | 0/0 | 0/0 | 0/0 | 0/0 | 0/0 | 0/0 | 15/10 | 0/0 | 0/0 | 17/15 | 11/11 | **79** |
| **12** | Mike Baldwin | USA | Bimota | 0/0 | 0/0 | 8/10 | 9/0 | 0/0 | 17/11 | 0/0 | 10/9 | 0/0 | 0/0 | 0/0 | **74** |
| **13** | Patrick Igoa | F | Kawasaki | 6/6 | 3/1 | 0/0 | 0/0 | 4/11 | 0/0 | 0/0 | 7/7 | 9/10 | 0/0 | 0/0 | **64** |
| **14** | Aaron Slight | NZ | Kawasaki | 0/0 | 0/0 | 0/0 | 0/0 | 0/0 | 0/0 | 7/9 | 0/0 | 0/0 | 11/10 | 17/0 | **54** |
| **15** | Alex Vieira | F | Honda | 0/0 | 0/0 | 0/0 | 0/0 | 20/8 | 0/0 | 0/0 | 15/8 | 0/0 | 0/0 | 0/0 | **51** |
| **16** | Malcolm Campbell | AUS | Honda | 0/0 | 1/4 | 0/0 | 0/0 | 9/10 | 0/0 | 1/6 | 0/0 | 0/0 | 0/0 | 0/17 | **48** |
| **17** | Davide Tardozzi | I | Bimota | 0/0 | 0/0 | 0/3 | 0/5 | 0/4 | 7/8 | 4/0 | 0/0 | 8/0 | 0/0 | 0/0 | **39** |
| **18** | Rueben McMurter | CDN | Honda | 0/0 | 0/0 | 15/17 | 6/0 | 0/0 | 0/0 | 0/0 | 0/0 | 0/0 | 0/0 | 0/0 | **38** |
| **19** | Roger Burnett | GB | Honda | 17/15 | 0/0 | 0/0 | 0/0 | 0/0 | 0/0 | 0/0 | 4/0 | 0/0 | 0/0 | 0/0 | **36** |
| **20** | Takahiro Sohwa | J | Kawasaki | 0/0 | 0/0 | 0/0 | 0/0 | 0/0 | 0/0 | 0/11 | 0/0 | 0/0 | 2/9 | 7/7 | **36** |
| **21** | Doug Polen | USA | Suzuki | 0/0 | 0/0 | 0/0 | 0/0 | 0/0 | 0/0 | 20/13 | 0/0 | 0/0 | 0/0 | 0/0 | **33** |
| **22** | Rob McElnea | GB | Yamaha | 0/0 | 13/15 | 0/0 | 0/0 | 0/0 | 0/0 | 0/0 | 0/0 | 0/0 | 0/0 | 0/0 | **28** |
| **23** | Ernst Gschwender | D | Suzuki | 5/9 | 8/0 | 0/0 | 0/0 | 1/5 | 0/0 | 0/0 | 0/0 | 0/0 | 0/0 | 0/0 | **28** |
| **24** | Kenichiro Iwahashi | J | Honda | 0/0 | 0/0 | 0/0 | 0/0 | 0/0 | 0/0 | 10/17 | 0/0 | 0/0 | 0/0 | 0/0 | **27** |
| **25** | Andreas Hofmann | D | Honda | 7/0 | 5/2 | 0/0 | 0/0 | 0/0 | 0/0 | 0/0 | 9/4 | 0/0 | 0/0 | 0/0 | **27** |
| **26** | Massimo Broccoli | I | Ducati | 0/0 | 0/0 | 0/0 | 0/0 | 13/13 | 0/0 | 0/0 | 0/0 | 0/0 | 0/0 | 0/0 | **26** |
| **27** | Rene Rasmussen | DK | Suzuki | 0/2 | 4/8 | 3/1 | 0/0 | 0/0 | 4/4 | 0/0 | 0/0 | 0/0 | 0/0 | 0/0 | **26** |
| **28** | Peter Goddard | AUS | Yamaha | 0/0 | 0/0 | 0/0 | 0/0 | 0/0 | 0/0 | 0/0 | 0/0 | 0/0 | 20/0 | 0/0 | **20** |
| **29** | Steve Crevier | CDN | Yamaha | 0/0 | 0/0 | 11/0 | 0/0 | 0/0 | 0/0 | 0/0 | 0/0 | 0/0 | 0/4 | 5/0 | **20** |
| **30** | Tommy Douglas | CDN | Yamaha | 0/0 | 0/0 | 6/0 | 7/7 | 0/0 | 0/0 | 0/0 | 0/0 | 0/0 | 0/0 | 0/0 | **20** |
| **31** | Tatsura Arata | J | Yamaha | 0/0 | 0/0 | 0/0 | 0/0 | 0/0 | 0/0 | 11/7 | 0/0 | 0/0 | 0/0 | 0/0 | **18** |
| **32** | Peter Rubatto | D | Bimota | 3/8 | 0/6 | 0/0 | 0/0 | 0/0 | 0/0 | 0/0 | 0/0 | 0/0 | 0/0 | 0/0 | **17** |
| **33** | Fabio Biliotti | I | Bimota | 0/0 | 0/0 | 0/4 | 5/8 | 0/0 | 0/0 | 0/0 | 0/0 | 0/0 | 0/0 | 0/0 | **17** |
| **34** | Brian Morrison | GB | Honda | 8/0 | 6/3 | 0/0 | 0/0 | 0/0 | 0/0 | 0/0 | 0/0 | 0/0 | 0/0 | 0/0 | **17** |
| **35** | Gary Goodfellow | CDN | Suzuki | 0/0 | 0/0 | 0/6 | 4/6 | 0/0 | 0/0 | 0/0 | 0/0 | 0/0 | 0/0 | 0/0 | **16** |
| **36** | Jean-Yves Mounier | F | Yamaha | 0/0 | 0/0 | 0/0 | 0/0 | 0/0 | 10/3 | 0/0 | 2/0 | 0/0 | 0/0 | 0/0 | **15** |
| **37** | Rene Bongers | AUS | Yamaha | 0/0 | 0/0 | 0/0 | 0/0 | 0/0 | 0/0 | 0/0 | 0/0 | 0/0 | 7/8 | 0/0 | **15** |
| **38** | Pierre Bolle | F | Kawasaki | 0/0 | 0/0 | 0/0 | 0/0 | 2/0 | 5/7 | 0/0 | 0/0 | 0/0 | 0/0 | 0/0 | **14** |
| **39** | Paul Iddon | GB | Yamaha | 0/4 | 0/0 | 4/0 | 0/0 | 0/0 | 0/6 | 0/0 | 0/0 | 0/0 | 0/0 | 0/0 | **14** |
| **40** | Andy McGladdery | GB | Honda | 0/5 | 0/0 | 0/0 | 0/0 | 0/0 | 0/0 | 0/0 | 0/0 | 0/0 | 0/2 | 2/5 | **14** |
| **41** | Steve Hislop | GB | Honda | 0/13 | 0/0 | 0/0 | 0/0 | 0/0 | 0/0 | 0/0 | 0/0 | 0/0 | 0/0 | 0/0 | **13** |
| **42** | Marco Lucchinelli | I | Ducati | 0/0 | 0/0 | 0/0 | 0/0 | 0/0 | 0/0 | 0/0 | 0/6 | 7/0 | 0/0 | 0/0 | **13** |
| **43** | Marino Fabbri | I | Bimota | 0/0 | 0/0 | 1/0 | 3/2 | 0/0 | 0/0 | 0/0 | 0/0 | 0/7 | 0/0 | 0/0 | **13** |
| **44** | Carl Fogarty | GB | Honda | 9/3 | 0/0 | 0/0 | 0/0 | 0/0 | 0/0 | 0/0 | 0/0 | 0/0 | 0/0 | 0/0 | **12** |
| **45** | Simon Crafar | NZ | Yamaha | 0/0 | 0/0 | 0/0 | 0/0 | 0/0 | 0/0 | 0/0 | 0/0 | 0/0 | 0/0 | 6/6 | **12** |
| **46** | Thierry Crine | F | Suzuki | 11/0 | 0/0 | 0/0 | 0/0 | 0/0 | 0/0 | 0/0 | 0/0 | 0/0 | 0/0 | 0/0 | **11** |
| **47** | Mark Phillips | GB | Yamaha | 4/7 | 0/0 | 0/0 | 0/0 | 0/0 | 0/0 | 0/0 | 0/0 | 0/0 | 0/0 | 0/0 | **11** |
| **48** | Andrew Stroud | NZ | Yamaha | 0/0 | 0/0 | 0/0 | 0/0 | 0/0 | 0/0 | 0/0 | 0/0 | 0/0 | 0/0 | 0/10 | **10** |
| **49** | Christophe Bouheben | F | Kawasaki | 1/0 | 0/0 | 0/0 | 0/0 | 0/0 | 0/0 | 0/0 | 0/1 | 0/8 | 0/0 | 0/0 | **10** |
| **50** | Giorgio Bontempi | I | Honda | 0/0 | 0/0 | 0/0 | 0/0 | 0/0 | 0/0 | 0/0 | 0/0 | 6/4 | 0/0 | 0/0 | **10** |
| **51** | Arnad Harmati | H | Honda | 0/0 | 0/0 | 0/0 | 0/0 | 0/0 | 0/0 | 0/0 | 0/0 | 5/5 | 0/0 | 0/0 | **10** |

# 1989 Superbike World Championship

| | | | | GB | H | CDN | USA | A | F | J | D | I | AUS | NZ | PTS |
|---|---|---|---|---|---|---|---|---|---|---|---|---|---|---|---|
| **52** | Bodo Schmidt | D | Yamaha | 0/0 | 0/0 | 0/0 | 0/0 | 0/0 | 0/0 | 0/0 | 5/5 | 0/0 | 0/0 | 0/0 | **10** |
| **53** | Shoichi Tsukamoto | J | Kawasaki | 0/0 | 0/0 | 0/0 | 0/0 | 0/0 | 0/0 | 5/5 | 0/0 | 0/0 | 0/0 | 0/0 | **10** |
| **54** | Glenn Williams | NZ | Ducati | 0/0 | 0/0 | 0/0 | 0/0 | 0/0 | 0/0 | 0/0 | 0/0 | 0/0 | 0/0 | 8/0 | **8** |
| **55** | Scott Doohan | AUS | Bimota | 0/0 | 0/0 | 0/0 | 0/0 | 0/0 | 0/0 | 2/0 | 0/0 | 0/0 | 6/0 | 0/0 | **8** |
| **56** | Matthew Blair | AUS | Suzuki | 0/0 | 0/0 | 0/0 | 0/0 | 0/0 | 0/0 | 0/0 | 0/0 | 0/0 | 0/7 | 0/0 | **7** |
| **57** | Maurice Coq | F | Yamaha | 0/0 | 0/0 | 0/0 | 0/0 | 0/0 | 2/5 | 0/0 | 0/0 | 0/0 | 0/0 | 0/0 | **7** |
| **58** | Michel Mercier | CDN | Suzuki | 0/0 | 0/0 | 5/2 | 0/0 | 0/0 | 0/0 | 0/0 | 0/0 | 0/0 | 0/0 | 0/0 | **7** |
| **59** | Eddie Kattenberg | NZ | Suzuki | 0/0 | 0/0 | 0/0 | 0/0 | 0/0 | 0/0 | 0/0 | 0/0 | 0/0 | 0/0 | 4/3 | **7** |
| **60** | Rene Delaby | B | Honda | 0/1 | 0/0 | 0/0 | 2/3 | 0/1 | 0/0 | 0/0 | 0/0 | 0/0 | 0/0 | 0/0 | **7** |
| **61** | Graeme Morris | AUS | Ducati | 0/0 | 0/0 | 0/0 | 0/0 | 0/0 | 0/0 | 0/0 | 0/0 | 0/0 | 0/6 | 0/0 | **6** |
| **62** | Marco Dall'Aglio | I | Yamaha | 0/0 | 0/0 | 0/0 | 0/0 | 0/0 | 0/0 | 0/0 | 0/0 | 0/6 | 0/0 | 0/0 | **6** |
| **63** | Naoto Abe | J | Honda | 0/0 | 0/0 | 0/0 | 0/0 | 0/0 | 0/0 | 6/0 | 0/0 | 0/0 | 0/0 | 0/0 | **6** |
| **64** | Corrado Manici | I | Ducati | 0/0 | 0/0 | 0/0 | 0/0 | 0/0 | 0/0 | 0/0 | 0/0 | 4/2 | 0/0 | 0/0 | **6** |
| **65** | Aldeo Presciutti | I | Yamaha | 0/0 | 0/0 | 0/0 | 0/0 | 0/0 | 0/0 | 0/0 | 0/0 | 3/3 | 0/0 | 0/0 | **6** |
| **66** | Robert Scolyer | AUS | Honda | 0/0 | 0/0 | 0/0 | 0/0 | 0/0 | 0/0 | 0/0 | 0/0 | 0/0 | 0/5 | 0/0 | **5** |
| **67** | Russell Josiah | NZ | Kawasaki | 0/0 | 0/0 | 0/0 | 0/0 | 0/0 | 0/0 | 0/0 | 0/0 | 0/0 | 0/0 | 3/2 | **5** |
| **68** | Edwin Weibel | CH | Kawasaki | 0/0 | 0/0 | 0/0 | 0/0 | 0/0 | 0/2 | 0/0 | 3/0 | 0/0 | 0/0 | 0/0 | **5** |
| **69** | Grant Ramage | NZ | Suzuki | 0/0 | 0/0 | 0/0 | 0/0 | 0/0 | 0/0 | 0/0 | 0/0 | 0/0 | 0/0 | 0/4 | **4** |
| **70** | Doug Chandler | USA | Kawasaki | 0/0 | 0/0 | 0/0 | 0/4 | 0/0 | 0/0 | 0/0 | 0/0 | 0/0 | 0/0 | 0/0 | **4** |
| **71** | John Richards | AUS | Yamaha | 0/0 | 0/0 | 0/0 | 0/0 | 0/0 | 0/0 | 0/0 | 0/0 | 0/0 | 3/1 | 0/0 | **4** |
| **72** | Andy Roberts | AUS | Yamaha | 0/0 | 0/0 | 0/0 | 0/0 | 0/0 | 0/0 | 0/0 | 0/0 | 0/0 | 0/3 | 0/0 | **3** |
| **73** | Udo Mark | D | Yamaha | 0/0 | 0/0 | 0/0 | 0/0 | 0/0 | 0/0 | 0/0 | 0/3 | 0/0 | 0/0 | 0/0 | **3** |
| **74** | Jeffrey de Vries | NL | Yamaha | 0/0 | 0/0 | 0/0 | 0/0 | 0/0 | 3/0 | 0/0 | 0/0 | 0/0 | 0/0 | 0/0 | **3** |
| **75** | Mauro Ricci | I | Kawasaki | 0/0 | 0/0 | 0/0 | 0/0 | 0/3 | 0/0 | 0/0 | 0/0 | 0/0 | 0/0 | 0/0 | **3** |
| **76** | Wolfgang von Muralt | CH | Honda | 0/0 | 0/0 | 2/0 | 0/1 | 0/0 | 0/0 | 0/0 | 0/0 | 0/0 | 0/0 | 0/0 | **3** |
| **77** | Silvano Richetti | I | Honda | 0/0 | 0/0 | 0/0 | 0/0 | 0/0 | 0/0 | 0/0 | 0/0 | 2/0 | 0/0 | 0/0 | **2** |
| **78** | Koichi Kobayashi | J | Honda | 0/0 | 0/0 | 0/0 | 0/0 | 0/0 | 0/0 | 0/2 | 0/0 | 0/0 | 0/0 | 0/0 | **2** |
| **79** | Herve Moineau | F | Suzuki | 2/0 | 0/0 | 0/0 | 0/0 | 0/0 | 0/0 | 0/0 | 0/0 | 0/0 | 0/0 | 0/0 | **2** |
| **80** | Scott Williams | NZ | Yamaha | 0/0 | 0/0 | 0/0 | 0/0 | 0/0 | 0/0 | 0/0 | 0/0 | 0/0 | 0/0 | 0/1 | **1** |
| **81** | Brian Bernard | NZ | Suzuki | 0/0 | 0/0 | 0/0 | 0/0 | 0/0 | 0/0 | 0/0 | 0/0 | 0/0 | 0/0 | 1/0 | **1** |
| **82** | Steven Martin | AUS | Suzuki | 0/0 | 0/0 | 0/0 | 0/0 | 0/0 | 0/0 | 0/0 | 0/0 | 0/0 | 1/0 | 0/0 | **1** |
| **83** | Oscar La Ferla | I | Yamaha | 0/0 | 0/0 | 0/0 | 0/0 | 0/0 | 0/0 | 0/0 | 0/0 | 0/1 | 0/0 | 0/0 | **1** |
| **84** | Adriano Narducci | I | Honda | 0/0 | 0/0 | 0/0 | 0/0 | 0/0 | 0/0 | 0/0 | 0/0 | 1/0 | 0/0 | 0/0 | **1** |
| **85** | Graeme McGregor | AUS | Honda | 0/0 | 0/0 | 0/0 | 0/0 | 0/0 | 0/0 | 0/0 | 1/0 | 0/0 | 0/0 | 0/0 | **1** |
| **86** | Koh Takeuchi | J | Yamaha | 0/0 | 0/0 | 0/0 | 0/0 | 0/0 | 0/0 | 0/1 | 0/0 | 0/0 | 0/0 | 0/0 | **1** |
| **87** | Eric Delcamp | F | Yamaha | 0/0 | 0/0 | 0/0 | 0/0 | 0/0 | 0/1 | 0/0 | 0/0 | 0/0 | 0/0 | 0/0 | **1** |
| **88** | Didier Hamdi | F | Yamaha | 0/0 | 0/0 | 0/0 | 0/0 | 0/0 | 1/0 | 0/0 | 0/0 | 0/0 | 0/0 | 0/0 | **1** |
| **89** | Russell Scott | USA | Suzuki | 0/0 | 0/0 | 0/0 | 1/0 | 0/0 | 0/0 | 0/0 | 0/0 | 0/0 | 0/0 | 0/0 | **1** |

# 1989 Manufacturers World Championship

| | | GB | H | CDN | USA | A | F | J | D | I | AUS | NZ | PTS |
|---|---|---|---|---|---|---|---|---|---|---|---|---|---|
| **1** | Honda | 17/15 | 20/20 | 20/17 | 15/17 | 20/20 | 20/15 | 10/17 | 17/15 | 20/17 | 8/13 | 15/20 | **368** |
| **2** | Yamaha | 20/17 | 15/17 | 11/13 | 17/13 | 11/17 | 13/9 | 17/20 | 13/11 | 15/13 | 20/20 | 20/13 | **335** |
| **3** | Ducati | 0/0 | 17/11 | 17/9 | 20/20 | 17/13 | 15/17 | 3/0 | 20/20 | 13/20 | 13/17 | 8/0 | **270** |
| **4** | Bimota | 3/20 | 7/13 | 8/20 | 9/8 | 3/4 | 17/20 | 13/15 | 10/17 | 8/7 | 6/0 | 0/0 | **208** |
| **5** | Kawasaki | 6/6 | 3/1 | 0/0 | 0/4 | 4/11 | 5/7 | 15/11 | 7/7 | 9/10 | 17/15 | 17/11 | **166** |
| **6** | Suzuki | 11/9 | 8/8 | 5/6 | 4/6 | 1/5 | 4/4 | 20/13 | 0/0 | 0/0 | 1/7 | 4/4 | **120** |